The Biosphere

The Biosphere

Edited by
Emmett Gonzalenz

Larsen & Keller
www.larsen-keller.com

The Biosphere
Edited by Emmett Gonzalenz
ISBN: 978-1-63549-049-7 (Hardback)

Larsen & Keller
Published by Larsen and Keller Education,
5 Penn Plaza,
19th Floor,
New York, NY 10001, USA

Cataloging-in-Publication Data

The biosphere / edited by Emmett Gonzalenz.
p. cm.
Includes bibliographical references and index.
ISBN 978-1-63549-049-7
1. Biosphere. 2. Geobiology. 3. Ecology. 4. Ecosystem management.
I. Gonzalenz, Emmett.
QH343.4 .B56 2017
577--dc23

The publisher's policy is to use permanent paper from mills that operate a sustainable forestry policy. Furthermore, the publisher ensures that the text paper and cover boards used have met acceptable environmental accreditation standards.

Printed and bound in China.

For more information regarding Larsen and Keller Education and its products, please visit the publisher's website www.larsen-keller.com

Table of Contents

Preface

This book elucidates the concepts and innovative models around prospective developments with respect to biosphere. It provides deep insights into this field. Biosphere refers to the collection of all the different ecosystems present on earth. It is a global ecological system which encompasses all the living beings and the process that relate them with each other and with the various natural elements like atmosphere, lithosphere, hydrosphere and geosphere. This text attempts to understand the multiple branches that fall under the discipline of biosphere and how such concepts have practical applications. Such selected concepts that redefine this field have been presented in it. This textbook will serve as a reference to a broad spectrum of readers.

A detailed account of the significant topics covered in this book is provided below:

Chapter 1- The sum of all ecosystems is known as biosphere. The biosphere is the global ecological system that supports life on earth. It started evolving 3.5 billion years ago. This chapter will provide an integrated understanding of biosphere.

Chapter 2- The mass of water that is found on or under the surface of Earth is known as hydrosphere. The gasses that cover the Earth form the atmosphere; it protects life by absorbing all ultraviolet radiation. The other components of biosphere that have been explained in this text are lithosphere, cryosphere, geosphere etc. This text strategically encompasses and incorporates the major components and key concepts of biosphere, providing a complete understanding.

Chapter 3- A community of living organisms is known as an ecosystem. The components that make up life on Earth can be divided into biotic and abiotic components; these are linked together through energy flows and natural cycles. This section is an overview of the subject matter incorporating all the major aspects of ecosystem.

Chapter 4- The study of the connections between organisms and their respective environments is known as ecology. This field involves biology, geography and Earth science. Some of the aspects of ecology elucidated in this chapter are biogenesis, abiogenesis, human ecology, biome, ecological niche and biomass. The topics explained in this section serve as a source to understand the theory of ecology.

Chapter 5- Biodiversity refers to all life on Earth. It is a measure of all organisms that exist in different ecosystems. The other aspects considered are environmental protection, habitat conservation, conservation biology and conservation movement.

The aim of this chapter is to explore all the aspects of biosphere. These topics are crucial for a complete understanding of the subject.

I would like to make a special mention of my publisher who considered me worthy of this opportunity and also supported me throughout the process. I would also like to thank the editing team at the back-end who extended their help whenever required.

Editor

1

Introduction to Biosphere

The sum of all ecosystems is known as biosphere. The biosphere is the global ecological system that supports life on earth. It started evolving 3.5 billion years ago. This chapter will provide an integrated understanding of biosphere.

Biosphere

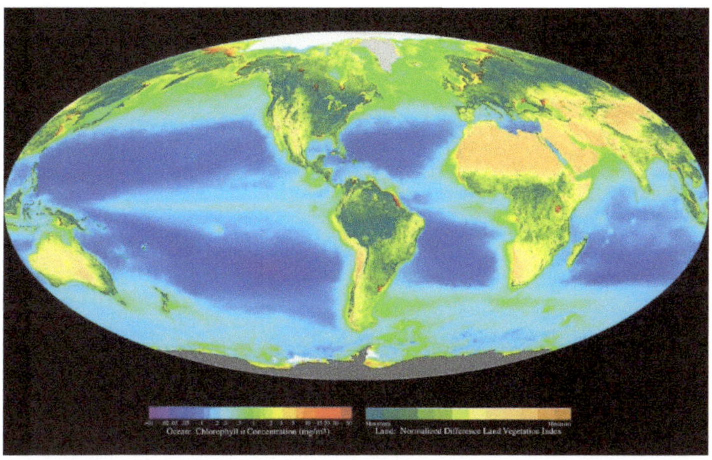

A false-color composite of global oceanic and terrestrial photoautotroph abundance, from September 1997 to August 2000. Provided by the SeaWiFS Project, NASA/Goddard Space Flight Center and ORBIMAGE.

The biosphere is the global sum of all ecosystems. It can also be termed as the zone of life on Earth, a closed system (apart from solar and cosmic radiation and heat from the interior of the Earth), and largely self-regulating. By the most general biophysiological definition, the biosphere is the global ecological system integrating all living beings and their relationships, including their interaction with the elements of the lithosphere, geosphere, hydrosphere, and atmosphere. The biosphere is postulated to have evolved, beginning with a process of biopoesis (life created naturally from non-living matter, such as simple organic compounds) or biogenesis (life created from living matter), at least some 3.5 billion years ago. The earliest evidence for life on Earth includes biogenic graphite found in 3.7 billion-year-old metasedimentary rocks from Western Greenland and microbial mat fossils found in 3.48 billion-year-old sandstone from Western Australia. More recently, in 2015, "remains of biotic life" were found in 4.1 billion-year-old rocks in Western Australia. According to one of the researchers, "If life arose relatively quickly on Earth ... then it cou ld be common in the universe."

In a general sense, biospheres are any closed, self-regulating systems containing ecosystems. This includes artificial biospheres such as Biosphere 2 and BIOS-3, and potentially ones on other planets or moons.

Origin and Use of the Term

A beach scene on Earth, simultaneously showing the lithosphere (ground), hydrosphere (ocean) and atmosphere (air)

The term "biosphere" was coined by geologist Eduard Suess in 1875, which he defined as the place on Earth's surface where life dwells.

While the concept has a geological origin, it is an indication of the effect of both Charles Darwin and Matthew F. Maury on the Earth sciences. The biosphere's ecological context comes from the 1920s, preceding the 1935 introduction of the term "ecosystem" by Sir Arthur Tansley. Vernadsky defined ecology as the science of the biosphere. It is an interdisciplinary concept for integrating astronomy, geophysics, meteorology, biogeography, evolution, geology, geochemistry, hydrology and, generally speaking, all life and Earth sciences.

Narrow Definition

Geochemists define the biosphere as being the total sum of living organisms (the "biomass" or "biota" as referred to by biologists and ecologists). In this sense, the biosphere is but one of four separate components of the geochemical model, the other three being *lithosphere*, *hydrosphere*, and *atmosphere*. The word *ecosphere*, coined during the 1960s, encompasses both biological and physical components of the planet.

The Second International Conference on Closed Life Systems defined *biospherics* as the science and technology of analogs and models of Earth's biosphere; i.e., artificial Earth-like biospheres. Others may include the creation of artificial non-Earth biospheres—for example, human-centered biospheres or a native Martian biosphere—as part of the topic of biospherics.

Extent of Earth's Biosphere

Water covers 71% of the Earth's surface. Image is the Blue Marble photographed from Apollo 17.

Every part of the planet, from the polar ice caps to the equator, features life of some kind. Recent advances in microbiology have demonstrated that microbes live deep beneath the Earth's terrestrial surface, and that the total mass of microbial life in so-called "uninhabitable zones" may, in biomass, exceed all animal and plant life on the surface. The actual thickness of the biosphere on earth is difficult to measure. Birds typically fly at altitudes as high as 1,800 m (5,900 ft; 1.1 mi) and fish live as much as 8,372 m (27,467 ft; 5.202 mi) underwater in the Puerto Rico Trench.

There are more extreme examples for life on the planet: Rüppell's vulture has been found at altitudes of 11,300 m (37,100 ft; 7.0 mi); bar-headed geese migrate at altitudes of at least 8,300 m (27,200 ft; 5.2 mi); yaks live at elevations as high as 5,400 m (17,700 ft; 3.4 mi) above sea level; mountain goats live up to 3,050 m (10,010 ft; 1.90 mi). Herbivorous animals at these elevations depend on lichens, grasses, and herbs.

Microscopic organisms live in every part of the biosphere, including soil, hot springs, "seven miles deep" in the ocean, "40 miles high" in the atmosphere and inside rocks far down within the Earth's crust. Microorganisms, under certain test conditions, have been observed to thrive in the vacuum of outer space. The total amount of soil and subsurface bacterial carbon is estimated as 5×10^{17} g, or the "weight of the United Kingdom". The mass of prokaryote microorganisms — which includes bacteria and archaea, but not the nucleated eukaryote microorganisms — may be as much as 0.8 trillion tons of carbon (of the total biosphere mass, estimated at between 1 and 4 trillion tons). Barophilic marine microbes have been found at more than a depth of 10,000 m (33,000 ft; 6.2 mi) in the Mariana Trench, the deepest spot in the Earth's oceans. In fact, single-celled life forms have been found in the deepest part of the Mariana Trench, by the Challenger Deep, at depths of 11,034 m (36,201 ft; 6.856 mi). Other researchers reported related studies that microorganisms thrive inside rocks up to 580 m (1,900 ft; 0.36 mi) below the sea floor under 2,590 m (8,500 ft; 1.61 mi) of ocean off the coast of the northwestern United States, as well as 2,400 m (7,900 ft; 1.5 mi) beneath the seabed off Japan. Culturable thermophilic microbes have been extracted from cores drilled more than 5,000 m (16,000 ft; 3.1 mi) into the Earth's crust in Sweden, from rocks between 65–75 °C (149–167 °F). Temperature increases with increasing depth into the Earth's crust. The rate at which the temperature increases depends on many factors, including type of crust (continental vs. oceanic), rock type, geographic location, etc. The greatest known temperature at which microbial life can exist is 122 °C (252 °F) (*Methanopyrus kandleri* Strain 116), and it is likely that the limit of life in the "deep biosphere" is defined by temperature rather than absolute depth. On 20 August 2014, scientists confirmed the existence of microorganisms living 800 m (2,600 ft; 0.50 mi) below the ice of Antarctica. According to one researcher, "You can find microbes everywhere — they're extremely adaptable to conditions, and survive wherever they are."

Our biosphere is divided into a number of biomes, inhabited by fairly similar flora and fauna. On land, biomes are separated primarily by latitude. Terrestrial biomes lying within the Arctic and Antarctic Circles are relatively barren of plant and animal life, while most of the more populous biomes lie near the equator.

Specific Biospheres

For this list, if a word is followed by a number, it is usually referring to a specific system or number. Thus:

- Biosphere 1, the planet Earth.
- Biosphere 2, laboratory in Arizona, United States, which contains 3.15 acres (13,000 m^2) of closed ecosystem.
- BIOS-3, a closed ecosystem at the Institute of Biophysics in Krasnoyarsk, Siberia, in what was then the Soviet Union.
- Biosphere J (CEEF, Closed Ecology Experiment Facilities), an experiment in Japan.

Extraterrestrial Biospheres

No biospheres have been detected beyond the Earth; therefore, the existence of extraterrestrial biospheres remains hypothetical. The rare Earth hypothesis suggests they should be very rare, save ones composed of microbial life only. On the other hand, Earth analogs may be quite numerous, at least in the Milky Way galaxy. Given limited understanding of abiogenesis, it is currently unknown what percentage of these planets actually develop biospheres.

It is also possible that artificial biospheres will be created during the future, for example on Mars. The process of creating an uncontained system that mimics the function of Earth's biosphere is called terraforming.

Biosphere 2

Biosphere 2 is an Earth systems science research facility located in Oracle, Arizona. It has been owned by the University of Arizona since 2011. Its mission is to serve as a center for research, outreach, teaching, and lifelong learning about Earth, its living systems, and its place in the universe. It is a 3.14-acre (1.27-hectare) structure originally built to be an artificial, materially closed ecological system, or vivarium. It remains the largest closed system ever created.

Biosphere 2 was originally meant to demonstrate the viability of closed ecological systems to support and maintain human life, defining mission one as eight humans for two years. Additionally, it served to explore the web of interactions within life systems in a structure with five areas based on biomes, and an agricultural area and human living and working space to study the interactions between humans, farming, and technology with the rest of nature. It also explored the use of closed biospheres in space colonization, and allowed the study and manipulation of a biosphere without harming Earth's. Its five biome areas were a 1,900 square meter rainforest, an 850 square meter ocean with a coral reef, a 450 square meter mangrove wetlands, a 1,300 square meter savannah grassland, a 1,400 square meter fog desert, a 2,500 square meter agricultural system, a human habitat, and a below-ground infrastructure. Heating and cooling water circulated through independent piping systems and passive solar input through the glass space frame panels covering most of the facility, and electrical power was supplied into Biosphere 2 from an onsite natural gas energy center.

Biosphere 2 was only used twice for its original intended purposes as a closed-system experiment: once from 1991 to 1993, and the second time from March to September 1994. Both attempts, though heavily publicized, ran into problems including low amounts of food and oxygen, die-offs of many animal and plant species, squabbling among the resident scientists and management issues.

In June 1994, during the middle of the second experiment, Space Biosphere Ventures dissolved, and the structure was left in limbo. It was purchased in 1995 by Columbia University, who used it to run experiments until 2005. It then looked in danger of being demolished to make way for housing and retail stores, but was taken over for research by the University of Arizona in 2007; the University of Arizona assumed full ownership of the structure in 2011.

Planning and Construction

Biosphere 2 was originally constructed between 1987 and 1991 by Space Biosphere Ventures, a joint venture whose principal officers were John P. Allen, inventor and Executive Director, and Margret Augustine, CEO; Marie Harding, Vice-President of Finance; Abigail Alling, Vice President of Research; and Norberto Alvarez-Romo, Vice President of Mission Control. Project funding came primarily from the joint venture's financial partner, Ed Bass's Decisions Investment. The project cost US$200 million from 1985 to 2007.

It was named "Biosphere 2" because it was meant to be the second fully self-sufficient biosphere, after the Earth itself.

Location

Biosphere 2 sits on a sprawling 40-acre (16-hectare) science campus that is open to the public

The glass facility is located in Oracle, Arizona at the base of the Santa Catalina Mountains, half an hour outside Tucson. It is elevated 4,000 feet (1,200 m) above sea level.

First Mission

The first closed mission lasted from September 26, 1991 to September 26, 1993. The crew were: medical doctor and researcher Roy Walford, Jane Poynter, Taber MacCallum, Mark Nelson, Sally Silverstone, Abigail Alling (who replaced Silke Schneider), Mark Van Thillo, and Linda Leigh.

The agricultural system produced 83% of the total diet, which included crops of bananas, papayas, sweet potatoes, beets, peanuts, lablab and cowpea beans, rice, and wheat. No toxic chemicals could be used, since they would impact health. During the first year the eight inhabitants reported continual hunger. During the second year, the crew produced over a ton more food, average caloric intake increased, and they regained some weight lost during the first year.

They consumed the same low-calorie, nutrient-dense diet which Roy Walford had studied in his research on extending lifespan through diet. Medical markers indicated the health of the crew during the two years was excellent. They showed the same improvement in health indices such as lowering of blood cholesterol, blood pressure, enhancement of immune system. They lost an average of 16% of their pre-entry body weight before stabilizing and regaining some weight during their second year. Subsequent studies showed that the biospherians' metabolism became more efficient at extracting nutrients from their food as an adaptation to the low-calorie, high nutrient diet.

Some of the domestic animals that were planned for the agricultural area during the first mission included four pygmy goats and one billy goat from the plateau region of Nigeria, 35 hens and three roosters (a mix of Indian jungle fowl (*Gallus gallus*), Japanese silky bantam, and a hybrid of these), two sows and one boar pig (feral), as well as tilapia fish grown in a rice and azolla pond system originating millennia ago in China.

A strategy of "species-packing" was practiced to ensure that food webs and ecological function could be maintained if some species did not survive. The fog desert area became more chaparral due to condensation from the space frame. The savannah was seasonally active; its biomass was cut and stored by the crew as part of their management of carbon dioxide. Rainforest pioneer species grew rapidly, but trees there and in the savannah suffered from etiolation and weakness caused by lack of stress wood, normally created in response to winds in natural conditions. Corals reproduced in the ocean area, and crew helped maintain ocean system health by hand-harvesting algae from the corals, manipulating calcium carbonate and pH levels to prevent the ocean becoming too acidic, and by installing an improved protein skimmer to supplement the algae turf scrubber system originally installed to remove excess nutrients. The mangrove area developed rapidly but with less understory than a typical wetland possibly because of reduced light levels.

Biosphere 2 suffered from CO_2 levels that "fluctuated wildly," and most of the vertebrate species and all of the pollinating insects died. Insect pests, like cockroaches, boomed. In practice, ants, a companion to one of the tree species (Cecropia) in the rainforest, had been introduced. By 1993 the tramp ant species *Paratrechina longicornis*, local to the area had been unintentionally sealed in and had come to dominate.

Challenges

Among the problems and miscalculations encountered in the first mission were overstocked fish dying and clogging filtration systems, unanticipated condensation making the "desert" too wet, population explosions of greenhouse ants and cockroaches, and morning glories overgrowing the "rainforest", blocking out other plants. In addition, construction itself was a challenge; for example, it was difficult to manipulate the bodies of water to have waves and tidal changes.

There was controversy when the public learned that the project had allowed an injured member to leave and return, carrying new material inside. The team claimed the only new supplies brought in were plastic bags, but others accused them of bringing food and other items. More criticism was raised when it was learned that, likewise, the project had been pumping oxygen inside, to make up for a failure in the balance of the system that resulted in the amount of oxygen steadily declining.

The oxygen inside the facility, which began at 20.9%, fell at a steady pace and after 16 months was down to 14.5%. This is equivalent to the oxygen availability at an elevation of 4,080 meters (13,400 ft). Since some biospherians were starting to have symptoms like sleep apnea and fatigue, Walford and the medical team decided to boost oxygen with injections in January and August 1993.

Daily fluctuation of carbon dioxide dynamics was typically 600 ppm because of the strong drawdown during sunlight hours by plant photosynthesis, followed by a similar rise during the nighttime when system respiration dominated. As expected, there was also a strong seasonal signature to CO_2 levels, with wintertime levels as high as 4,000-4,500 ppm and summertime levels near 1,000 ppm. The crew worked to manage the CO_2 by occasionally turning on a CO_2 scrubber, activating and de-activating the desert and savannah through control of irrigation water, cutting and storing biomass to sequester carbon, and utilizing all potential planting areas with fast-growing species to increase system photosynthesis.

Many suspected the drop in oxygen was due to microbes in the soil. The soils were selected to have enough carbon to provide for the plants of the ecosystems to grow from infancy to maturity, a plant mass increase of perhaps 20 tons (18,000 kg). The release rate of that soil carbon as carbon dioxide by respiration of soil microbes was an unknown that the Biosphere 2 experiment was designed to reveal.

The respiration rate was faster than the photosynthesis (possibly in part due to relatively low light penetration through the glazed structure) resulting in a slow decrease of oxygen. A mystery accompanied the oxygen decline: the corresponding increase in carbon dioxide did not appear. This concealed the underlying process until an investigation by Jeff Severinghaus and Wallace Broecker of Columbia University's Lamont Doherty Earth Observatory using isotopic analysis showed that carbon dioxide was reacting with exposed concrete inside Biosphere 2 to form calcium carbonate, thereby sequestering both carbon and oxygen.

Second Mission

During the transition period between missions, extensive research and system improvements had been undertaken. Concrete was sealed to prevent uptake of carbon dioxide. The second mission began on March 6, 1994, with an announced run of ten months. Crew was Norberto Alvarez-Romo (Capt.), John Druitt, Matt Finn, Pascale Maslin, Charlotte Godfrey, Rodrigo Romo and Tilak Mahato. The second crew achieved complete sufficiency in food production.

On April 1, 1994 a severe dispute within the management team led to the ousting of the on-site management by federal marshals serving a restraining order, leaving management of the mission to the Bannon & Co. team from Beverly Hills, California.

At 3 am on April 5, 1994, Abigail Alling and Mark Van Thillo, members of the first crew, allegedly vandalized the project from outside, opening one double-airlock door and three single door emergency exits, leaving them open for approximately fifteen minutes. Five panes of glass were also broken. About 10% of the biosphere's air was exchanged with the outside during this time, according to Donella Meadows, who had a communication from Ms. Alling in which she explained that they wanted to give those inside the choice of continuing or leaving, as she didn't know what they had been told of the new situation.

Soon after that, the captain Norberto Alvarez-Romo (by then married to Margret Augustine) precipitously exited the Biosphere to assist the managerial process under the new circumstances. He was replaced by Bernd Zabel, who had been nominated as captain of the first mission but who was replaced at the last minute. Two months later, Matt Smith replaced Matt Finn.

The ownership and management company Space Biospheres Ventures was officially dissolved on June 1, 1994. This left the scientific and business management of the mission to the interim turnaround team, who had been contracted by the financial partner, Decisions Investment Co.

Mission 2 was ended prematurely on September 6, 1994.

Columbia University

After a successful turnaround by Bannon & Co. in December 1995 the Biosphere 2 owners transferred management to Columbia University of New York City which embarked on a successful eight-year run at the Biosphere 2 campus. Columbia ran Biosphere 2 as a research site and campus until 2003. Subsequently, management reverted to the owners.

In 1996, Columbia University changed the virtually airtight, materially closed structure designed for closed system research, to a "flow-through" system, and halted closed system research. They manipulated carbon dioxide levels for global warming research, and injected desired amounts of carbon dioxide, venting as needed. During Columbia's tenure, Columbia students would often spend one semester at the site.

Site Sold

In January 2005, Decisions Investments Corporation, owner of Biosphere 2, announced that the project's 1,600-acre (650 ha) campus was for sale. They preferred a research use to be found for the complex but were not excluding buyers with different intentions, such as universities, churches, resorts, and spas. In June 2007 the site was sold for $50 million to CDO Ranching & Development, L.P. 1,500 houses and a resort hotel were planned, but the main structure was still to be available for research and educational use.

Acquisition by University of Arizona

On June 26, 2007, the University of Arizona announced it would take over research at the Biosphere 2. The announcement ended fears that the structure would be demolished. University officials said private gifts and grants enabled them to cover research and operating costs for three years with the possibility of extending funding for ten years. It was extended for ten years, and is now engaged in research projects including research into the terrestrial water cycle and how it relates to ecology, atmospheric science, soil geochemistry, and climate change. In June 2011, the University announced that it would assume full ownership of Biosphere 2, effective July 1.

Biosphere 2, 2015

CDO Ranching & Development donated the land, Biosphere buildings and several other support and administrative buildings. The Philecology Foundation (a nonprofit research foundation founded by Ed Bass) pledged US$20 million for the ongoing science and operations.

Engineering

The above-ground physical structure of Biosphere 2 was made of steel tubing and high-performance glass and steel frames. The frame and glazing materials were designed and made to specification by a firm run by a one-time associate of Buckminster Fuller, Peter Jon Pearce (Pearce Structures, Inc.). The window seals and structures had to be designed to be almost perfectly airtight, such that the air exchange would be extremely slow, to avoid damage to the experimental results.

Biosphere 2 from the inside. Seen here are the Savanna *(foreground)* and Ocean *(background)* sections.

The fog desert section of Biosphere in 2005

During the day, the heat from the sun caused the air inside to expand and during the night it cooled and contracted. To avoid having to deal with the huge forces that maintaining a constant volume would create, the structure had large diaphragms kept in domes called "lungs".

Since opening a window was not an option, the structure also required huge air conditioners to control the temperature and avoid killing the plants within. For every unit of solar energy that entered the structure, the air conditioners would expend approximately three times as much energy to cool the habitat back down.

Science

A special issue of the Ecological Engineering journal edited by Marino and Howard T.

Odum (1999), published as "Biosphere 2: Research Past and Present" (Elsevier, 1999) represents the most comprehensive assemblage of collected papers and findings from Biosphere 2. The papers range from calibrated models that describe the system metabolism, hydrologic balance, and heat and humidity, to papers that describe rainforest, mangrove, ocean, and agronomic system development in this carbon dioxide-rich environment.

Praise and Criticism

One view of Biosphere 2 was that it was "the most exciting scientific project to be undertaken in the U.S. since President John F. Kennedy launched us toward the moon". Others called it "New Age drivel masquerading as science". John Allen and Roy Walford did have mainstream credentials. John Allen held a degree in Metallurgical-Mining Engineering from the Colorado School of Mines, and an MBA from the Harvard Business School. Roy Walford received his doctorate of medicine from the University of Chicago and taught at UCLA as a Professor of Pathology for 35 years. Shortly after leaving Biosphere 2's first mission, Mark Nelson obtained his Ph. D in 1998 under Professor H.T. Odum in Ecological Engineering.

Questioning the credentials of the participants (despite the contribution in the preparation phase of Biosphere 2 of worldwide top-level scientists and among others the Russian Academy of Sciences), Marc Cooper wrote that "the group that built, conceived, and directs the Biosphere project is not a group of high-tech researchers on the cutting edge of science but a clique of recycled theater performers that evolved out of an authoritarian—and decidedly non-scientific—personality cult". He was referring to the Synergia Ranch in New Mexico, where indeed many of the Biospherians did practice theater under John Allen's leadership, and began to develop the ideas behind Biosphere 2. However, the original Biosphere 2 Science Advisory Committee, chaired by Tom Lovejoy of the Smithsonian Institution, in the report of August 1992 reported: "The committee is in agreement that the conception and construction of Biosphere 2 were acts of vision and courage. The scale of Biosphere 2 is unique and Biosphere 2 is already providing unexpected scientific results not possible through other means (notably the documented, unexpected decline in atmospheric oxygen levels.) Biosphere 2 will make important scientific contributions in the fields of biogeochemical cycling, the ecology of closed ecological systems, and restoration ecology." Furthermore, Columbia University assembled outside scientists to evaluate the potential of the facility, and concluded the following: "A group of world-class scientists got together and decided the Biosphere 2 facility is an exceptional laboratory for addressing critical questions relative to the future of Earth and its environment."

One of their own scientific consultants came to be critical of the enterprise, too. Dr. Ghillean Prance, director of the Royal Botanical Gardens in Kew, designed the rainforest biome inside the Biosphere. Although he later recanted, acknowledging the unique scope of this experiment and contributed to its success as a consultant, in a 1983 interview (8 years before the start of the experiment), Prance said, "I was attracted to the Institute of Ecotechnics because funds for research were being cut and the institute seemed to have a lot of money which it was willing to spend freely. Along with others, I was ill-used. Their interest in science is not genuine. They seem to have some sort of secret agenda, they seem to be guided by some sort of religious or philosophical system."

The Ethnosphere: Psychology and Conflict

The crew quarters

Much of the evidence for isolated human groups comes from psychological studies of scientists overwintering in Antarctic research stations. The study of this phenomenon is "confined environment psychology", and according to Jane Poynter not nearly enough of it was brought to bear on Biosphere 2.

Before the first closure mission was half over, the group had split into two factions and people who had been intimate friends had become implacable enemies, barely on speaking terms.

The faction inside the bubble came from a rift between the joint venture partners on how the science should proceed, as biospherics or as specialist ecosystem studies (perceived as reductionist and precisely contrary to the raison d'être of the experiment). The faction that included Poynter felt strongly that they should be making formal proposals for research for the Science Advisory Committee to evaluate. The other faction included Abigail Alling, the titular director of research inside the bubble, and who sided with John Allen in blocking that move. On February 14, the entire SAC resigned. *Time* Magazine wrote:

Now, the veneer of credibility, already bruised by allegations of tamper-prone data, secret food caches and smuggled supplies, has cracked ... the two-year experiment in self-sufficiency is starting to look less like science and more like a $150 million stunt.

Undoubtedly the lack of oxygen and the calorie-restricted, nutrient-dense diet contributed to low morale. The Alling faction feared that the Poynter group were prepared to go so far as to import food, if it meant making them fitter to carry out research projects. They considered that would be a project failure by definition.

In November 1992, the hungry Biospherians began eating emergency food supplies that had not been grown inside the bubble. Poynter made Chris Helms, PR Director for the enterprise, aware of this. She was promptly dismissed by Margret Augustine, CEO of Space Biospheres Ventures, and told to come out of the biosphere. This order was, however, never carried out. Poynter writes that she simply decided to stay put, correctly reasoning that the order could not be enforced without effectively terminating the closure.

BIOS-3

BIOS-3 is a closed ecosystem at the Institute of Biophysics in Krasnoyarsk, Russia.

Its construction began in 1965, and was completed in 1972. BIOS-3 consists of a 315 m3 underground steel structure suitable for up to three persons, and was initially used for developing closed ecosystems capable of supporting humans. It was divided into 4 compartments — one of which is a crew area. The crew area consists of 3 single-cabins, a galley, lavatory and control room. Initially one other compartment was an algal cultivator, and the other two 'phytrons' for growing wheat or vegetables. The plants growing in the two 'phytrons' contributed approximately 25% of the air filtering in the compound. Later the algal cultivator was converted into a third phytron. A level of light comparable to sunlight was supplied in each of the 4 compartments by 20 kW xenon lamps, cooled by water jackets. The facility used 400 kW of electricity, supplied by a nearby hydroelectric power station.

Chlorella algae were used to recycle air breathed by humans, absorbing carbon dioxide and replenishing it with oxygen through photosynthesis. The algae were cultivated in stacked tanks under artificial light. To achieve a balance of oxygen and carbon dioxide, one human needed 8 m2 of exposed Chlorella. Air was purified of more complex organic compounds by heating to 600 °C in the presence of a catalyst. Water and nutrients were stored in advance and were also recycled. By 1968, system efficiency had reached 85% by recycling water. Dried meat was imported into the facility, and urine and feces were generally dried and stored, rather than being recycled.

BIOS-3 facilities were used to conduct 10 manned closure experiments with a one to three man crew. The longest experiment with a three-man crew lasted 180 days (in 1972-1973). The facilities were used for the tests at least until 1984.

In 1991, BIOS-3 became a part of the International Center for Closed Ecosystems, which was formed as a subdivision of Institute of Biophysics, Russian Academy of Sciences, Siberian Branch. Closed ecosystems research focusing on growing plants and recycling waste was resumed in 2005 in cooperation with European Space Agency.

Shadow Biosphere

A shadow biosphere is a hypothetical microbial biosphere of Earth that uses radically different biochemical and molecular processes than currently known life. Although life on Earth is relatively well-studied, the shadow biosphere may still remain unnoticed because the exploration of the microbial world targets primarily the biochemistry of the macro-organisms. The term was coined by Carol Cleland and Shelley Copley in 2005.

The Hypothesis

Steven A. Benner, Alonso Ricardo, and Matthew A. Carrigan, biochemists at the University of Florida, argued that if organisms based on RNA once existed, they might still be alive today, unnoticed because they don't contain ribosomes, which are usually used to detect living organisms. They suggest searching for them in environments that are low in sulfur, environments that are spatially constrained (for example, minerals with pores smaller than one micrometre), or environments that cycle between extreme hot and cold.

Other proposed candidates for a shadow biosphere include organisms using different suites of amino acids in their proteins or different molecular units (e.g., bases or sugars) in their nucleic acids, having a chirality opposite of ours, using some of the non-standard amino acids, or using arsenic instead of phosphorus. Carol Cleland, a philosopher of science at the University of Colorado (Boulder), argues that desert varnish, whose status as living or nonliving has been debated since the time of Darwin, should be investigated as a potential candidate for a shadow biosphere.

Criticism

The idea of a shadow biosphere is not widely accepted in biochemistry, and methods used by proponents and conclusions drawn from experiments that purport to show evidence of shadow biospheres have been criticized. For example, evidence that once seemed to support arsenic as a substitute for phosphorus in DNA could have resulted from lab or field contamination, and DNA that includes arsenic is chemically unstable.

References

- Campbell, Neil A.; Brad Williamson; Robin J. Heyden (2006). Biology: Exploring Life. Boston, Massachusetts: Pearson Prentice Hall. ISBN 0-13-250882-6.
- Möller, Detlev (December 2010). Chemistry of the Climate System. De Gruyter. pp. 118–119. ISBN 978-3-11-022835-9.
- Bebarta, Kailash Chandra (2011). Dictionary of Forestry and Wildlife Science. New Delhi: Concept Publishing Company. p. 45. ISBN 978-81-8069-719-7.
- Trembath-Reichert, Elizabeth (2005-02-04). "Biosphere 2 Now for Sale to Highest Bidder". Columbia Daily Spectator. Retrieved 2016-03-13.

- Borenstein, Seth (19 October 2015). "Hints of life on what was thought to be desolate early Earth". Excite. Yonkers, NY: Mindspark Interactive Network. Associated Press. Retrieved 2015-10-20.
- University of Georgia (25 August 1998). "First-Ever Scientific Estimate Of Total Bacteria On Earth Shows Far Greater Numbers Than Ever Known Before". Science Daily. Retrieved 10 November 2014.
- Morelle, Rebecca (15 December 2014). "Microbes discovered by deepest marine drill analysed". BBC News. Retrieved 15 December 2014.
- Fox, Douglas (20 August 2014). "Lakes under the ice: Antarctica's secret garden". Nature. 512: 244–246. Bibcode:2014Natur.512..244F. doi:10.1038/512244a. Retrieved 21 August 2014.
- Mack, Eric (20 August 2014). "Life Confirmed Under Antarctic Ice; Is Space Next?". Forbes. Retrieved 21 August 2014.
- Borenstein, Seth (13 November 2013). "Oldest fossil found: Meet your microbial mom". AP News. Retrieved 15 November 2013.
- Zimmer, Carl (3 October 2013). "Earth's Oxygen: A Mystery Easy to Take for Granted". New York Times. Retrieved 3 October 2013.
- Oskin, Becky (14 March 2013). "Intraterrestrials: Life Thrives in Ocean Floor". LiveScience. Retrieved 17 March 2013.
- Choi, Charles Q. (17 March 2013). "Microbes Thrive in Deepest Spot on Earth". LiveScience. Retrieved 17 March 2013.
- "Biosphere 2 to Have a Permanent Home With the UA". Office of University Communications, The University of Arizona. 2011-06-27. Retrieved 2011-06-27.

2

Components of Biosphere

The mass of water that is found on or under the surface of Earth is known as hydrosphere. The gasses that cover the Earth form the atmosphere; it protects life by absorbing all ultraviolet radiation. The other components of biosphere that have been explained in this text are lithosphere, cryosphere, geosphere etc. This text strategically encompasses and incorporates the major components and key concepts of biosphere, providing a complete understanding.

Geosphere

There are several conflicting definitions for geosphere.

The geosphere may be taken as the collective name for the lithosphere, the hydrosphere, the cryosphere, and the atmosphere.

In Aristotelian physics, the term was applied to four spherical *natural places*, concentrically nested around the center of the Earth, as described in the lectures *Physica* and *Meteorologica*. They were believed to explain the motions of the four *terrestrial elements: Earth*, *Water*, *Air* and *Fire*.

In modern texts and in Earth system science, geosphere refers to the solid parts of the Earth; it is used along with atmosphere, hydrosphere, and biosphere to describe the systems of the Earth (the interaction of these systems with the magnetosphere is sometimes listed). In that context, sometimes the term lithosphere is used instead of geosphere or solid Earth. The lithosphere, however, only refers to the uppermost layers of the solid Earth (oceanic and continental crustal rocks and uppermost mantle).

Since space exploration began, it has been observed that the extent of the ionosphere or plasmasphere is highly variable, and often much larger than previously appreciated, at times extending to the boundaries of the Earth's magnetosphere or geomagnetosphere. This highly variable outer boundary of *geogenic* matter has been referred to as the "geopause," to suggest the relative scarcity of such matter beyond it, where the solar wind dominates. wowza

Hydrosphere

The hydrosphere is the combined mass of water found on, under, and above the surface of a planet.

It has been estimated that there are 1386 million cubic kilometres of water on Earth.

This includes water in liquid and frozen forms in groundwater, oceans, lakes and streams. Saltwater accounts for 97.5% of this amount. Fresh water accounts for only 2.5%. Of this fresh water, 68.7% is in the form of ice and permanent snow cover in the Arctic, the Antarctic, and mountain glaciers. 29.9% is in the form of fresh groundwater. Only 0.26% of the fresh water on Earth is in easily accessible lakes, reservoirs and river systems. The total mass of the Earth's hydrosphere is about 1.4×10^{18} tonnes, which is about 0.023% of Earth's total mass. About 20×10^{12} tonnes of this is in Earth's atmosphere (for practical purposes, 1 cubic metre of water weighs one tonne). Approximately 75% of Earth's surface, an area of some 361 million square kilometers (139.5 million square miles), is covered by ocean. The average salinity of Earth's oceans is about 35 grams of salt per kilogram of sea water (3.5%).

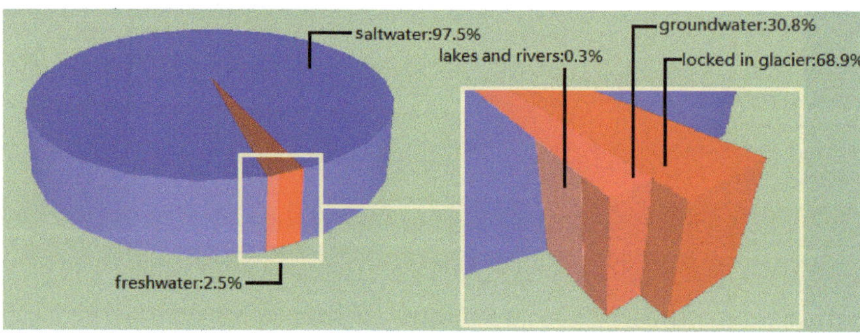

Water Cycle

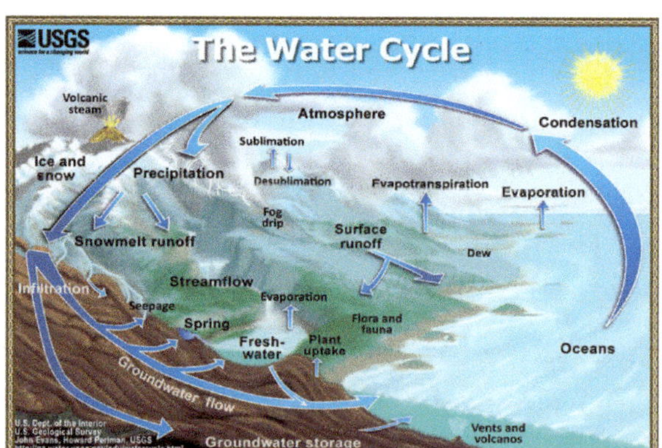

The hydrological cycle transfers water from one state or reservoir to another. Reservoirs include atmospheric moisture (snow, rain and clouds),streams, oceans, rivers, lakes, groundwater, subterranean aquifers, polar icecaps and saturated soil. Solar energy, in the form of heat and light (insolation), and gravity cause the transfer from one state to another over periods from hours to thousands of years. Most evaporation comes from the oceans and is returned to the earth as snow or rain (page 27). Sublimation refers to evaporation from snow and ice. Transpiration refers to the expiration of water through the minute pores or stomata of trees. Evapotranspiration is the term used by hydrologists in reference to the three processes together, transpiration, sublimation and evaporation.

In his book *Water*, Marq de Villiers described the hydrosphere as a closed system in which water exists. The hydrosphere is intricate, complex, interdependent, all-pervading and stable and "seems purpose-built for regulating life (de Villiers 2003:26)." De Villiers claimed that, "On earth, the total amount of water has almost certainly not changed since geological times: what we had then we still have. Water can be polluted, abused, and misused but it is neither created nor destroyed, it only migrates. There is no evidence that water vapor escapes into space (page 26)."

"Every year the turnover of water on Earth involves 577,000 km^3 of water. This is water that evaporates from the oceanic surface (502,800 km3) and from land (74,200 km^3). The same amount of water falls as atmospheric precipitation, 458,000 km^3 on the ocean and 119,000 km^3 on land. The difference between precipitation and evaporation from the land surface (119,000 - 74,200 = 44,800 km^3/year) represents the total runoff of the Earth's rivers (42,700 km^3/year) and direct groundwater runoff to the ocean (2100 km^3/year). These are the principal sources of fresh water to support life necessities and man's economic activities."

Water is a basic necessity of life. Since 2/3 of the Earth is covered by water, the Earth is also called the blue planet and the watery planet. Hydrosphere plays an important role in the existence of the atmosphere in its present form. Oceans are important in this regard. When the Earth was formed it had only a very thin atmosphere rich in hydrogen and helium similar to the present atmosphere of Mercury. Later the gases hydrogen and helium were expelled from the atmosphere. The gases and water vapor released as the Earth cooled became our present atmosphere. Other gases and water vapor released by volcanoes also entered the atmosphere. As the Earth cooled the water vapor in the atmosphere condensed and fell as rain. The atmosphere cooled further as atmospheric carbon dioxide dissolved in to rain water. In turn this further caused the water vapor to condense and fall as rain. This rain water filled the depressions on the Earth's surface and formed the oceans. It is estimated that this occurred about 4000 million years ago. The first life forms began in the oceans. These organisms did not breathe oxygen. Later, when cyanobacteria evolved, the process of conversion of carbon dioxide into food and oxygen began. As a result, our atmosphere has a distinctly different composition from that of the other planets; it is a fundamental requirement for life on Earth.

Recharging Reservoirs

According to Igor A. Shiklomanov, it takes 2500 years for the complete recharge and replenishment of oceanic waters, 10,000 years for permafrost and ice, 1500 years for deep groundwater and mountainous glaciers, 17 years in lakes and 16 days in rivers.

Specific Fresh Water Availability

"Specific water availability is the residual (after use) per capita quantity of fresh water." Fresh water resources are unevenly distributed in terms of space and time and can go from floods to water shortages within months in the same area. In 1998 76% of the total population had a specific water availability of less than 5.0 thousand m3 per year per capita. Already by 1998, 35% of the global population suffered "very low or catastrophically low water supplies" and Shiklomanov predicted that the situation would deteriorate in the twenty-first century with "most of the Earth's population will be living under the conditions of low or catastrophically low water supply" by 2025. There is only 2.5% of fresh water in the hydrosphere.

Atmosphere of Earth

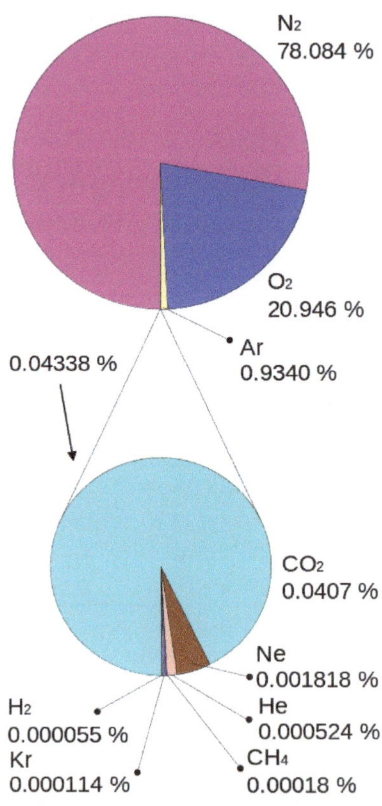

Composition of Earth's atmosphere by volume. The lower pie represents the trace gases that together compose about 0.038% of the atmosphere (0.043% with CO_2 at 2014 concentration). The numbers are from a variety of years (mainly 1987, with CO_2 and methane from 2009) and do not represent any single source.

Diagram of Earth's atmosphere (layers to scale). Distance from the surface to the top of the stratosphere is just under 1% of Earth's radius.

Blue light is scattered more than other wavelengths by the gases in the atmosphere, giving Earth a blue halo when seen from space onboard *ISS* at a height of 402–424 km

The atmosphere of Earth is the layer of gases, commonly known as **air**, that surrounds the planet Earth and is retained by Earth's gravity. The atmosphere protects life on Earth by absorbing ultraviolet solar radiation, warming the surface through heat retention (greenhouse effect), and reducing temperature extremes between day and night (the diurnal temperature variation).

By volume, dry air contains 78.09% nitrogen, 20.95% oxygen, 0.93% argon, 0.039% carbon dioxide, and small amounts of other gases. Air also contains a variable amount of water vapor, on average around 1% at sea level, and 0.4% over the entire atmosphere. Air content and atmospheric pressure vary at different layers, and air suitable for use in photosynthesis by terrestrial plants and breathing of terrestrial animals is found only in Earth's troposphere and in artificial atmospheres.

The atmosphere has a mass of about 5.15×10^{18} kg, three quarters of which is within about 11 km (6.8 mi; 36,000 ft) of the surface. The atmosphere becomes thinner and thinner with increasing altitude, with no definite boundary between the atmosphere and outer space. The Kármán line, at 100 km (62 mi), or 1.57% of Earth's radius, is often used as the border between the atmosphere and outer space. Atmospheric effects become noticeable during atmospheric reentry of spacecraft at an altitude of around 120 km (75 mi). Several layers can be distinguished in the atmosphere, based on characteristics such as temperature and composition.

The study of Earth's atmosphere and its processes is called atmospheric science (aerology). Early pioneers in the field include Léon Teisserenc de Bort and Richard Assmann.

Composition

The three major constituents of air, and therefore of Earth's atmosphere, are nitrogen, oxygen, and argon. Water vapor accounts for roughly 0.25% of the atmosphere by mass. The concentration of water vapor (a greenhouse gas) varies significantly from around 10 ppm by volume in the coldest portions of the atmosphere to as much as 5% by volume in hot, humid air masses, and concentrations of other atmospheric gases are typically quoted in terms of dry air (without water vapor). The remaining gases are often referred to as trace gases, among which are the greenhouse gases, principally carbon dioxide, methane, nitrous oxide, and ozone. Filtered air includes trace amounts of many other chemical compounds. Many substances of natural origin may be present in locally and seasonally variable small amounts as aerosols in an unfiltered air sample, including dust of mineral and organic composition, pollen and spores, sea spray, and volcanic ash. Various industrial pollutants also may be present as gases or aerosols, such as chlorine (elemental or in compounds), fluorine compounds and elemental mercury vapor. Sulfur compounds such as hydrogen sulfide and sulfur dioxide (SO_2) may be derived from natural sources or from industrial air pollution.

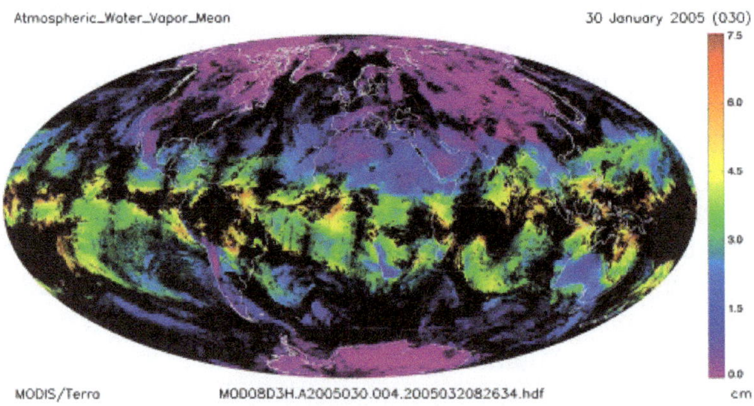

Mean atmospheric water vapor

Major constituents of dry air, by volume			
Gas		**Volume**[(A)]	
Name	Formula	in ppmv[(B)]	in %
Nitrogen	N_2	780,840	78.084
Oxygen	O_2	209,460	20.946
Argon	Ar	9,340	0.9340
Carbon dioxide	CO_2	397	0.0397
Neon	Ne	18.18	0.001818
Helium	He	5.24	0.000524
Methane	CH_4	1.79	0.000179
Not included in above dry atmosphere:			
Water vapor[(C)]	H_2O	10–50,000[(D)]	0.001%–5%[(D)]

notes:
(A) volume fraction is equal to mole fraction for ideal gas only
(B) ppmv: parts per million by volume
(C) Water vapor is about 0.25% *by mass* over full atmosphere
(D) Water vapor strongly varies locally

Structure of the Atmosphere

Principal Layers

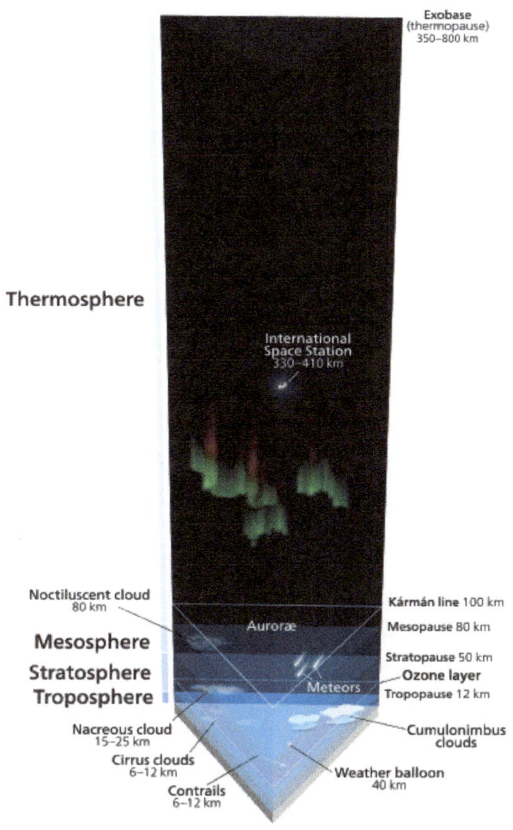

Earth's atmosphere Lower 4 layers of the atmosphere in 3 dimensions as seen diagonally from above the exobase. Layers drawn to scale, objects within the layers are not to scale. Aurorae shown here at the bottom of the thermosphere can actually form at any altitude in this atmospheric layer

In general, air pressure and density decrease with altitude in the atmosphere. However, temperature has a more complicated profile with altitude, and may remain relatively constant or even increase with altitude in some regions. Because the general pattern of the temperature/altitude profile is constant and measurable by means of instrumented balloon soundings, the temperature behavior provides a useful metric to distinguish atmospheric layers. In this way, Earth's atmosphere can be divided (called atmospheric stratification) into five main layers. Excluding the exosphere, Earth has four primary layers, which are the troposphere, stratosphere, mesosphere, and thermosphere. From highest to lowest, the five main layers are:

- Exosphere: 700 to 10,000 km (440 to 6,200 miles)
- Thermosphere: 80 to 700 km (50 to 440 miles)
- Mesosphere: 50 to 80 km (31 to 50 miles)
- Stratosphere: 12 to 50 km (7 to 31 miles)
- Troposphere: 0 to 12 km (0 to 7 miles)

Exosphere

The exosphere is the outermost layer of Earth's atmosphere (i.e. the upper limit of the atmosphere). It extends from the exobase, which is located at the top of the thermosphere at an altitude of about 700 km above sea level, to about 10,000 km (6,200 mi; 33,000,000 ft) where it merges into the solar wind.

This layer is mainly composed of extremely low densities of hydrogen, helium and several heavier molecules including nitrogen, oxygen and carbon dioxide closer to the exobase. The atoms and molecules are so far apart that they can travel hundreds of kilometers without colliding with one another. Thus, the exosphere no longer behaves like a gas, and the particles constantly escape into space. These free-moving particles follow ballistic trajectories and may migrate in and out of the magnetosphere or the solar wind.

The exosphere is located too far above Earth for any meteorological phenomena to be possible. However, the aurora borealis and aurora australis sometimes occur in the lower part of the exosphere, where they overlap into the thermosphere. The exosphere contains most of the satellites orbiting Earth.

Thermosphere

The thermosphere is the second-highest layer of Earth's atmosphere. It extends from the mesopause (which separates it from the mesosphere) at an altitude of about 80 km (50 mi; 260,000 ft) up to the thermopause at an altitude range of 500–1000 km (310–620 mi; 1,600,000–3,300,000 ft). The height of the thermopause varies considerably due to changes in solar activity. Because the thermopause lies at the lower boundary of the exosphere, it is also referred to as the exobase. The lower part of the thermosphere, from 80 to 550 kilometres (50 to 342 mi) above Earth's surface, contains the ionosphere.

The temperature of the thermosphere gradually increases with height. Unlike the stratosphere beneath it, wherein a temperature inversion is due to the absorption of radiation by ozone, the inversion in the thermosphere occurs due to the extremely low density of its molecules. The temperature of this layer can rise as high as 1500 °C (2700 °F), though the gas molecules are so far apart that its temperature in the usual sense is not very meaningful. The air is so rarefied that an individual molecule (of oxygen, for example) travels an average of 1 kilometre (0.62 mi; 3300 ft) between collisions with other molecules. Although the thermosphere has a high proportion of molecules with high energy, it would not feel hot to a human in direct contact, because its density is too low to conduct a significant amount of energy to or from the skin.

This layer is completely cloudless and free of water vapor. However non-hydrometeorological phenomena such as the aurora borealis and aurora australis are occasionally seen in the thermosphere. The International Space Station orbits in this layer, between 350 and 420 km (220 and 260 mi).

Mesosphere

The mesosphere is the third highest layer of Earth's atmosphere, occupying the region above the stratosphere and below the thermosphere. It extends from the stratopause at an altitude of about 50 km (31 mi; 160,000 ft) to the mesopause at 80–85 km (50–53 mi; 260,000–280,000 ft) above sea level.

Temperatures drop with increasing altitude to the mesopause that marks the top of this middle layer of the atmosphere. It is the coldest place on Earth and has an average temperature around −85 °C (−120 °F; 190 K).

Just below the mesopause, the air is so cold that even the very scarce water vapor at this altitude can be sublimated into polar-mesospheric noctilucent clouds. These are the highest clouds in the atmosphere and may be visible to the naked eye if sunlight reflects off them about an hour or two after sunset or a similar length of time before sunrise. They are most readily visible when the Sun is around 4 to 16 degrees below the horizon. A type of lightning referred to as either sprites or ELVES, occasionally form far above tropospheric thunderclouds. The mesosphere is also the layer where most meteors burn up upon atmospheric entrance. It is too high above Earth to be accessible to jet-powered aircraft and balloons, and too low to permit orbital spacecraft. The mesosphere is mainly accessed by sounding rockets and rocket-powered aircraft.

Stratosphere

The stratosphere is the second-lowest layer of Earth's atmosphere. It lies above the troposphere and is separated from it by the tropopause. This layer extends from the top of the troposphere at roughly 12 km (7.5 mi; 39,000 ft) above Earth's surface to the stratopause at an altitude of about 50 to 55 km (31 to 34 mi; 164,000 to 180,000 ft).

The atmospheric pressure at the top of the stratosphere is roughly 1/1000 the pressure at sea level. It contains the ozone layer, which is the part of Earth's atmosphere that contains relatively high concentrations of that gas. The stratosphere defines a layer in which temperatures rise with increasing altitude. This rise in temperature is caused by the absorption of ultraviolet radiation (UV) radiation from the Sun by the ozone layer, which restricts turbulence and mixing. Although the temperature may be −60 °C (−76 °F; 210 K) at the tropopause, the top of the stratosphere is much warmer, and may be near 0 °C.

The stratospheric temperature profile creates very stable atmospheric conditions, so the stratosphere lacks the weather-producing air turbulence that is so prevalent in the troposphere. Consequently, the stratosphere is almost completely free of clouds and other forms of weather. However, polar stratospheric or nacreous clouds are occasionally seen in the lower part of this layer of the atmosphere where the air is coldest. This is the highest layer that can be accessed by jet-powered aircraft.

Troposphere

The troposphere is the lowest layer of Earth's atmosphere. It extends from Earth's surface to an average height of about 12 km, although this altitude actually varies from about 9 km (30,000 ft) at the poles to 17 km (56,000 ft) at the equator, with some variation due to weather. The troposphere is bounded above by the tropopause, a boundary marked in most places by a temperature inversion (i.e. a layer of relatively warm air above a colder one), and in others by a zone which is isothermal with height.

Although variations do occur, the temperature usually declines with increasing altitude in the troposphere because the troposphere is mostly heated through energy transfer from the surface. Thus, the lowest part of the troposphere (i.e. Earth's surface) is typically the warmest section of the troposphere. The troposphere contains roughly 80% of the mass of Earth's atmosphere. The troposphere is denser than all its overlying atmospheric layers because a larger atmospheric weight sits on top of the troposphere and causes it to be most severely compressed. Fifty percent of the total mass of the atmosphere is located in the lower 5.6 km (18,000 ft) of the troposphere.

Space Shuttle Endeavour orbiting in the thermosphere. Because of the angle of the photo, it appears to straddle the stratosphere and mesosphere that actually lie more than 250 km below. The orange layer is the troposphere, which gives way to the whitish stratosphere and then the blue mesosphere

Nearly all atmospheric water vapor or moisture is found in the troposphere, so it is the layer where most of Earth's weather takes place. It has basically all the weather-associated cloud genus types generated by active wind circulation, although very tall cumulonimbus thunder clouds can penetrate the tropopause from below and rise into the lower part of the stratosphere. Most conventional aviation activity takes place in the troposphere, and it is the only layer that can be accessed by propeller-driven aircraft.

Other Layers

Within the five principal layers that are largely determined by temperature, several secondary layers may be distinguished by other properties:

- The ozone layer is contained within the stratosphere. In this layer ozone concentrations are about 2 to 8 parts per million, which is much higher than in the lower atmosphere but still very small compared to the main components of the atmosphere. It is mainly located in the lower portion of the stratosphere from about 15–35 km (9.3–21.7 mi; 49,000–115,000 ft), though the thickness varies seasonally and geographically. About 90% of the ozone in Earth's atmosphere is contained in the stratosphere.
- The ionosphere is a region of the atmosphere that is ionized by solar radiation. It is responsible for auroras. During daytime hours, it stretches from 50 to 1,000 km (31 to 621 mi; 160,000 to 3,280,000 ft) and includes the mesosphere, thermosphere, and parts of the exosphere. However, ionization in the mesosphere largely ceases during the night, so auroras are normally seen only in the thermosphere and lower exosphere. The ionosphere forms the inner edge of the magnetosphere. It has practical importance because it influences, for example, radio propagation on Earth.
- The homosphere and heterosphere are defined by whether the atmospheric gases are well mixed. The surface-based homosphere includes the troposphere, stratosphere, mesosphere, and the lowest part of the thermosphere, where the chemical composition of the atmosphere does not depend on molecular weight because the gases are mixed by turbulence. This relatively homogeneous layer ends at the *turbopause* found at about 100 km (62 mi; 330,000 ft), which places it about 20 km (12 mi; 66,000 ft) above the mesopause.

  Above this altitude lies the heterosphere, which includes the exosphere and most of the thermosphere. Here, the chemical composition varies with altitude. This is because the distance that particles can move without colliding with one another is large compared with the size of motions that cause mixing. This allows the gases to stratify by molecular weight, with the heavier ones, such as oxygen and nitrogen, present only near the bottom of the heterosphere. The upper part of the heterosphere is composed almost completely of hydrogen, the lightest element.

- The planetary boundary layer is the part of the troposphere that is closest to Earth's surface and is directly affected by it, mainly through turbulent diffusion. During the day the planetary boundary layer usually is well-mixed, whereas at night it becomes stably stratified with weak or intermittent mixing. The depth of the planetary boundary layer ranges from as little as about 100 meters on clear, calm nights to 3000 m or more during the afternoon in dry regions.

The average temperature of the atmosphere at Earth's surface is 14 °C (57 °F; 287 K) or 15 °C (59 °F; 288 K), depending on the reference.

Physical Properties

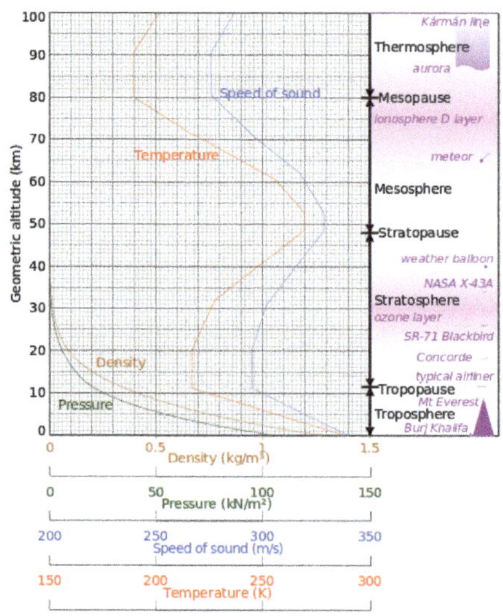

Comparison of the 1962 US Standard Atmosphere graph of geometric altitude against air density, pressure, the speed of sound and temperature with approximate altitudes of various objects.

Pressure and Thickness

The average atmospheric pressure at sea level is defined by the International Standard Atmosphere as 101325 pascals (760.00 Torr; 14.6959 psi; 760.00 mmHg). This is sometimes referred to as a unit of standard atmospheres (atm). Total atmospheric mass is 5.1480×10^{18} kg (1.135×10^{19} lb), about 2.5% less than would be inferred from the average sea level pressure and Earth's area of 51007.2 megahectares, this portion being displaced by Earth's mountainous terrain. Atmospheric pressure is the total weight of the air above unit area at the point where the pressure is measured. Thus air pressure varies with location and weather.

If the entire mass of the atmosphere had a uniform density from sea level, it would terminate abruptly at an altitude of 8.50 km (27,900 ft). It actually decreases exponentially with altitude, dropping by half every 5.6 km (18,000 ft) or by a factor of 1/e every 7.64 km (25,100 ft), the average scale height of the atmosphere below 70 km (43 mi; 230,000 ft). However, the atmosphere is more accurately modeled with a customized equation for each layer that takes gradients of temperature, molecular composition, solar radiation and gravity into account.

In summary, the mass of Earth's atmosphere is distributed approximately as follows:

- 50% is below 5.6 km (18,000 ft).
- 90% is below 16 km (52,000 ft).
- 99.99997% is below 100 km (62 mi; 330,000 ft), the Kármán line. By international convention, this marks the beginning of space where human travelers are considered astronauts.

By comparison, the summit of Mt. Everest is at 8,848 m (29,029 ft); commercial airliners typically cruise between 10 km (33,000 ft) and 13 km (43,000 ft) where the thinner air improves fuel economy; weather balloons reach 30.4 km (100,000 ft) and above; and the highest X-15 flight in 1963 reached 108.0 km (354,300 ft).

Even above the Kármán line, significant atmospheric effects such as auroras still occur. Meteors begin to glow in this region, though the larger ones may not burn up until they penetrate more deeply. The various layers of Earth's ionosphere, important to HF radio propagation, begin below 100 km and extend beyond 500 km. By comparison, the International Space Station and Space Shuttle typically orbit at 350–400 km, within the F-layer of the ionosphere where they encounter enough atmospheric drag to require reboosts every few months. Depending on solar activity, satellites can experience noticeable atmospheric drag at altitudes as high as 700–800 km.

Temperature and Speed of Sound

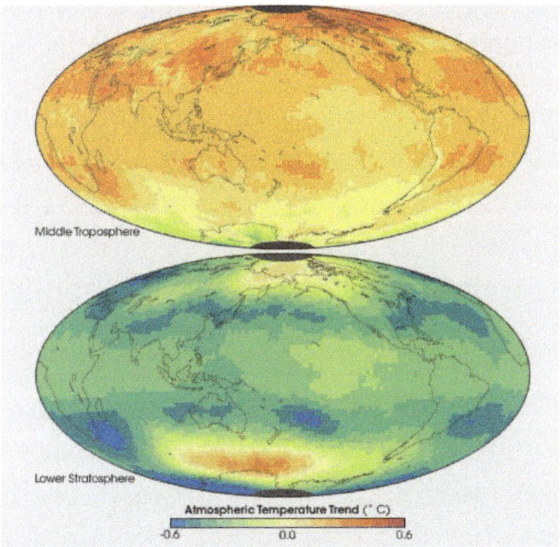

These images show temperature trends in two thick layers of the atmosphere as measured by a series of satellite-based instruments between January 1979 and December 2005. The measurements were taken by Microwave Sounding Units and Advanced Microwave Sounding Units flying on a series of National Oceanic and Atmospheric Administration (NOAA) weather satellites. The instruments record microwaves emitted from oxygen molecules in the atmosphere. Source:

The division of the atmosphere into layers mostly by reference to temperature is discussed above. Temperature decreases with altitude starting at sea level, but variations in this trend begin above 11 km, where the temperature stabilizes through a large vertical distance through the rest of the troposphere. In the stratosphere, starting above about 20 km, the temperature increases with height, due to heating within the ozone layer caused by capture of significant ultraviolet radiation from the Sun by the dioxygen and ozone gas in this region. Still another region of increasing temperature with altitude occurs at very high altitudes, in the aptly-named thermosphere above 90 km.

Because in an ideal gas of constant composition the speed of sound depends only on temperature and not on the gas pressure or density, the speed of sound in the atmosphere with altitude takes on the form of the complicated temperature profile, and does not mirror altitudinal changes in density or pressure.

Density and Mass

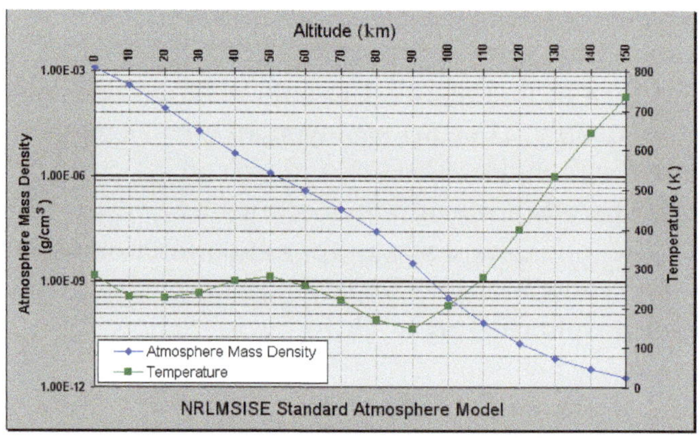

Temperature and mass density against altitude from the NRLMSISE-00 standard atmosphere model (the eight dotted lines in each "decade" are at the eight cubes 8, 27, 64, ..., 729)

The density of air at sea level is about 1.2 kg/m³ (1.2 g/L, 0.0012 g/cm³). Density is not measured directly but is calculated from measurements of temperature, pressure and humidity using the equation of state for air (a form of the ideal gas law). Atmospheric density decreases as the altitude increases. This variation can be approximately modeled using the barometric formula. More sophisticated models are used to predict orbital decay of satellites.

The average mass of the atmosphere is about 5 quadrillion (5×10^{15}) tonnes or 1/1,200,000 the mass of Earth. According to the American National Center for Atmospheric Research, "The total mean mass of the atmosphere is 5.1480×10^{18} kg with an annual range due to water vapor of 1.2 or 1.5×10^{15} kg, depending on whether surface pressure or water vapor data are used; somewhat smaller than the previous estimate. The mean mass of water vapor is estimated as 1.27×10^{16} kg and the dry air mass as $5.1352 \pm 0.0003 \times 10^{18}$ kg."

Optical Properties

Solar radiation (or sunlight) is the energy Earth receives from the Sun. Earth also emits radiation back into space, but at longer wavelengths that we cannot see. Part of the incoming and emitted radiation is absorbed or reflected by the atmosphere.

Scattering

When light passes through Earth's atmosphere, photons interact with it through *scattering*. If the light does not interact with the atmosphere, it is called *direct radiation* and is what you see if you were to look directly at the Sun. *Indirect radiation* is light that has been scattered in the atmosphere. For example, on an overcast day when you cannot see your shadow there is no direct radiation reaching you, it has all been scattered. As another example, due to a phenomenon called Rayleigh scattering, shorter (blue) wavelengths scatter more easily than longer (red) wavelengths. This is why the sky looks blue; you are seeing scattered blue light. This is also why sunsets are red. Because the Sun is close to the horizon, the Sun's rays pass through more atmosphere than normal to reach your eye. Much of the blue light has been scattered out, leaving the red light in a sunset.

Absorption

Different molecules absorb different wavelengths of radiation. For example, O_2 and O_3 absorb almost all wavelengths shorter than 300 nanometers. Water (H_2O) absorbs many wavelengths above 700 nm. When a molecule absorbs a photon, it increases the energy of the molecule. This heats the atmosphere, but the atmosphere also cools by emitting radiation, as discussed below.

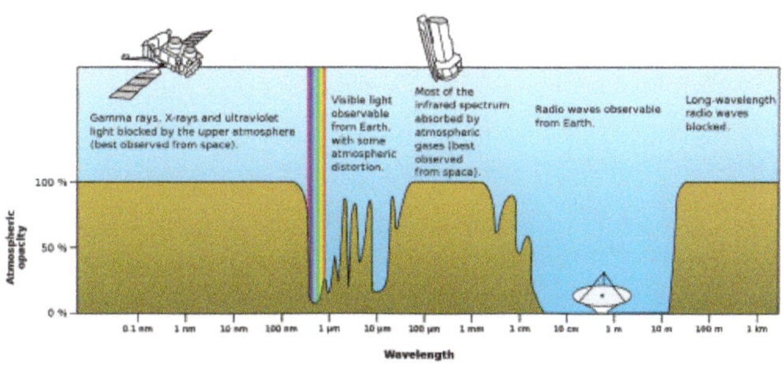

Rough plot of Earth's atmospheric transmittance (or opacity) to various wavelengths of electromagnetic radiation, including visible light.

The combined absorption spectra of the gases in the atmosphere leave "windows" of low opacity, allowing the transmission of only certain bands of light. The optical window runs from around 300 nm (ultraviolet-C) up into the range humans can see, the visible spectrum (commonly called light), at roughly 400–700 nm and continues to the infrared to around 1100 nm. There are also infrared and radio windows that transmit some infrared and radio waves at longer wavelengths. For example, the radio window runs from about one centimeter to about eleven-meter waves.

Emission

Emission is the opposite of absorption, it is when an object emits radiation. Objects tend to emit amounts and wavelengths of radiation depending on their "black body" emission curves, therefore hotter objects tend to emit more radiation, with shorter wavelengths. Colder objects emit less radiation, with longer wavelengths. For example, the Sun is approximately 6,000 K (5,730 °C; 10,340 °F), its radiation peaks near 500 nm, and is visible to the human eye. Earth is approximately 290 K (17 °C; 62 °F), so its radiation peaks near 10,000 nm, and is much too long to be visible to humans.

Because of its temperature, the atmosphere emits infrared radiation. For example, on clear nights Earth's surface cools down faster than on cloudy nights. This is because clouds (H_2O) are strong absorbers and emitters of infrared radiation. This is also why it becomes colder at night at higher elevations.

The greenhouse effect is directly related to this absorption and emission effect. Some gases in the atmosphere absorb and emit infrared radiation, but do not interact with sunlight in the visible spectrum. Common examples of these are CO_2 and H_2O.

Refractive Index

The refractive index of air is close to, but just greater than 1. Systematic variations in refractive index can lead to the bending of light rays over long optical paths. One example is that, under some circumstances, observers onboard ships can see other vessels just over the horizon because light is refracted in the same direction as the curvature of Earth's surface.

The refractive index of air depends on temperature, giving rise to refraction effects when the temperature gradient is large. An example of such effects is the mirage.

Circulation

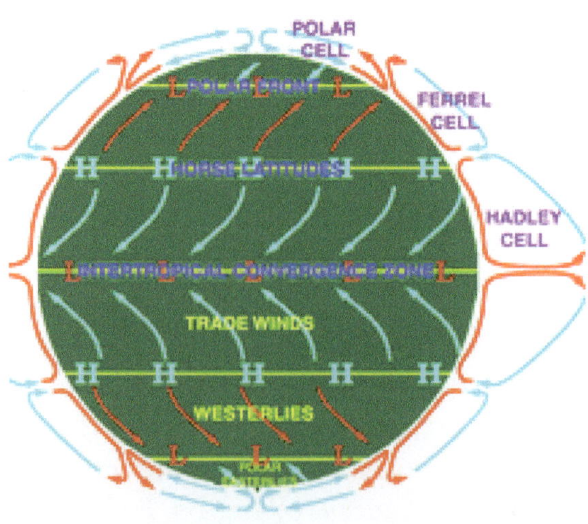

An idealised view of three large circulation cells.

Atmospheric circulation is the large-scale movement of air through the troposphere, and the means (with ocean circulation) by which heat is distributed around Earth. The large-scale structure of the atmospheric circulation varies from year to year, but the basic structure remains fairly constant because it is determined by Earth's rotation rate and the difference in solar radiation between the equator and poles.

Evolution of Earth's Atmosphere

Earliest Atmosphere

The first atmosphere would have consisted of gases in the solar nebula, primarily hydrogen. In addition, there would probably have been simple hydrides such as those now found in the gas giants (Jupiter and Saturn), notably water vapor, methane and ammonia. As the solar nebula dissipated, these gases would have escaped, partly driven off by the solar wind.

Second Atmosphere

Outgassing from volcanism, supplemented by gases produced during the late heavy bombardment of Earth by huge asteroids, produced the next atmosphere, consisting largely of nitrogen plus carbon dioxide and inert gases. A major part of carbon-dioxide emissions soon dissolved in water and built up carbonate sediments.

Researchers have found water-related sediments dating from as early as 3.8 billion years ago. About 3.4 billion years ago, nitrogen formed the major part of the then stable "second atmosphere". An influence of life has to be taken into account rather soon in the history of the atmosphere, because hints of early life-forms appear as early as 3.5 billion years ago. How Earth at that time maintained a climate warm enough for liquid water and life, if the early Sun put out 30% lower solar radiance than today, is a puzzle known as the "faint young Sun paradox".

The geological record however shows a continually relatively warm surface during the complete early temperature record of Earth - with the exception of one cold glacial phase about 2.4 billion years ago. In the late Archean eon an oxygen-containing atmosphere began to develop, apparently produced by photosynthesizing cyanobacteria, which have been found as stromatolite fossils from 2.7 billion years ago. The early basic carbon isotopy (isotope ratio proportions) very much approximates current conditions, suggesting that the fundamental features of the carbon cycle became established as early as 4 billion years ago.

Ancient sediments in the Republic of Gabon dating from between about 2,150 and 2,080 million years ago provide a record of Earth's dynamic oxygenation evolution. These fluctuations in oxygenation were likely driven by the Lomagundi carbon isotope excursion.

Third Atmosphere

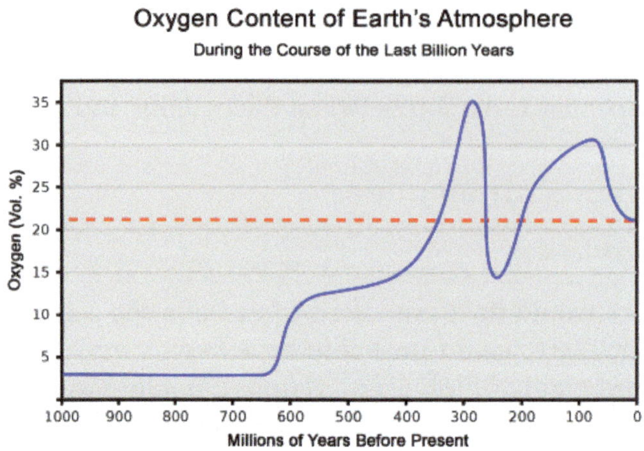

Oxygen content of the atmosphere over the last billion years. This diagram in more detail

The constant re-arrangement of continents by plate tectonics influences the long-term evolution of the atmosphere by transferring carbon dioxide to and from large continental carbonate stores. Free oxygen did not exist in the atmosphere until about 2.4 billion years ago during the Great Oxygenation Event and its appearance is indicated by the end of the banded iron formations. Before this time, any oxygen produced by photosynthesis was consumed by oxidation of reduced materials, notably iron. Molecules of free oxygen did not start to accumulate in the atmosphere until the rate of production of oxygen began to exceed the availability of reducing materials. This point signifies a shift from a reducing atmosphere to an oxidizing atmosphere. O_2 showed major variations until reaching a steady state of more than 15% by the end of the Precambrian. The following time span from 541 million years ago to the present day is the Phanerozoic eon, during the earliest period of which, the Cambrian, oxygen-requiring metazoan life forms began to appear.

The amount of oxygen in the atmosphere has fluctuated over the last 600 million years, reaching a peak of about 30% around 280 million years ago, significantly higher than today's 21%. Two main processes govern changes in the atmosphere: Plants use carbon dioxide from the atmosphere, releasing oxygen. Breakdown of pyrite and volcanic eruptions release sulfur into the atmosphere, which oxidizes and hence reduces the amount of oxygen in the atmosphere. However, volcanic eruptions also release carbon dioxide, which plants can convert to oxygen. The exact cause of the variation of the amount of oxygen in the atmosphere is not known. Periods with much oxygen in the atmosphere are associated with rapid development of animals. Today's atmosphere contains 21% oxygen, which is high enough for this rapid development of animals.

The scientific consensus is that the anthropogenic greenhouse gases currently accumulating in the atmosphere are the main cause of global warming.

Air Pollution

Air pollution is the introduction into the atmosphere of chemicals, particulate matter or biological materials that cause harm or discomfort to organisms. Stratospheric ozone depletion is caused by air pollution, chiefly from chlorofluorocarbons and other ozone-depleting substances.

Images from Space

On October 19, 2015 NASA started a website containing daily images of the full sunlit side of Earth. The images are taken from the Deep Space Climate Observatory (DSCOVR) and show Earth as it rotates during a day.

The geomagnetic storms cause beautiful displays of aurora across the atmosphere.

Limb view, of Earth's atmosphere. Colors roughly denote the layers of the atmosphere.

This image shows the Moon at the centre, with the limb of Earth near the bottom transitioning into the orange-colored troposphere. The troposphere ends abruptly at the tropopause, which appears in the image as the sharp boundary between the orange- and blue-colored atmosphere. The silvery-blue noctilucent clouds extend far above Earth's troposphere.

Earth's atmosphere backlit by the Sun in an eclipse observed from deep space onboard Apollo 12 in 1969.

Lithosphere

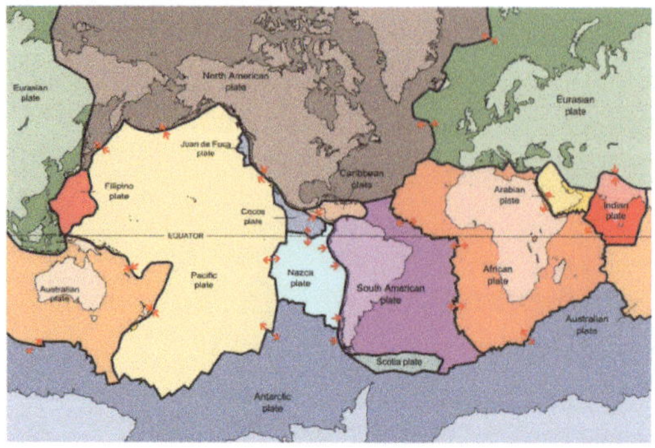

The tectonic plates of the lithosphere on Earth

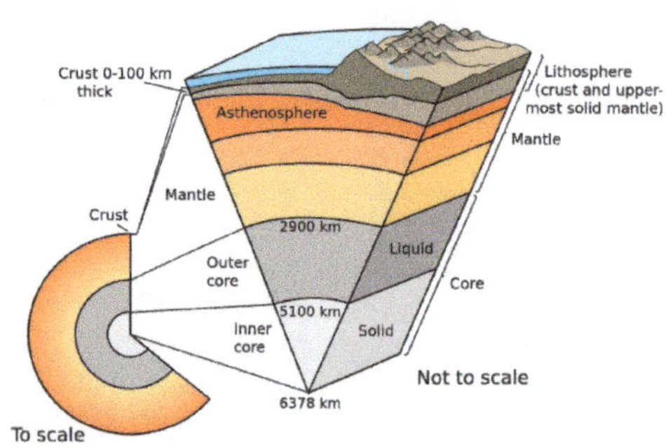

Earth cutaway from core to crust, the lithosphere comprising the crust and lithospheric mantle (detail not to scale)

A lithosphere is the rigid, outermost shell of a terrestrial-type planet or natural satellite that is defined by its rigid mechanical properties. On Earth, it is composed of the crust and the portion of the upper mantle that behaves elastically on time scales of thousands of years or greater. The outermost shell of a rocky planet, the crust, is defined on the basis of its chemistry and mineralogy.

Earth's Lithosphere

Earth's lithosphere includes the crust and the uppermost mantle, which constitute the hard and rigid outer layer of the Earth. The lithosphere is subdivided into tectonic plates. The uppermost part of the lithosphere that chemically reacts to the atmosphere, hydrosphere and biosphere through the soil forming process is called the pedosphere. The lithosphere is underlain by the asthenosphere which is the weaker, hotter, and deeper part of the upper mantle. The boundary between the lithosphere and the underlying asthenosphere is known as the Lithosphere-Asthenosphere boundary and is defined by a difference in response to stress: the lithosphere remains rigid for very long periods of geologic time in which it deforms elastically and through brittle failure, while the asthenosphere deforms viscously and accommodates strain through plastic deformation. The study of past and current formations of landscapes is called geomorphology.

History

The concept of the lithosphere as Earth's strong outer layer was described by A.E.H. Love in his 1911 monograph "Some problems of Geodynamics" and further developed by Joseph Barrell, who wrote a series of papers about the concept and introduced the term "lithosphere". The concept was based on the presence of significant gravity anomalies over continental crust, from which he inferred that there must exist a strong upper layer (which he called the lithosphere) above a weaker layer which could flow (which he called the asthenosphere). These ideas were expanded by Reginald Aldworth Daly in 1940 with his seminal work "Strength and Structure of the Earth" and have been broadly accepted by geologists and geophysicists. Although these ideas about lithosphere and asthenosphere were developed long before plate tectonic theory was articulated in the 1960s, the concepts that a strong lithosphere exists and that this rests on a weak asthenosphere are essential to that theory.

Types

There are two types of lithosphere:

- Oceanic lithosphere, which is associated with oceanic crust and exists in the ocean basins (mean density of about 2.9 grams per cubic centimeter)
- Continental lithosphere, which is associated with continental crust (mean density of about 2.7 grams per cubic centimeter)

The thickness of the lithosphere is considered to be the depth to the isotherm associated with the transition between brittle and viscous behavior. The temperature at which olivine begins to deform viscously (~1000 °C) is often used to set this isotherm because olivine is generally the weakest mineral in the upper mantle. Oceanic lithosphere is typically about 50–140 km thick (but beneath the mid-ocean ridges is no thicker than the crust), while continental lithosphere has a range in thickness from about 40 km to perhaps 280 km; the upper ~30 to ~50 km of typical continental lithosphere is crust. The mantle part of the lithosphere consists largely of peridotite. The crust is distinguished from the upper mantle by the change in chemical composition that takes place at the Moho discontinuity.

Oceanic Lithosphere

Oceanic lithosphere consists mainly of mafic crust and ultramafic mantle (peridotite) and is denser than continental lithosphere, for which the mantle is associated with crust made of felsic rocks. Oceanic lithosphere thickens as it ages and moves away from the mid-ocean ridge. This thickening occurs by conductive cooling, which converts hot asthenosphere into lithospheric mantle and causes the oceanic lithosphere to become increasingly thick and dense with age. The thickness of the mantle part of the oceanic lithosphere can be approximated as a thermal boundary layer that thickens as the square root of time.

$$h \sim 2\sqrt{\kappa t}$$

Here, κ is the thickness of the oceanic mantle lithosphere, is the thermal diffusivity (approximately 10^{-6} m^2/s) for silicate rocks, and t is the age of the given part of the lithosphere. The age is often equal to L/V, where L is the distance from the spreading centre of mid-oceanic ridge, and V is velocity of the lithospheric plate.

Oceanic lithosphere is less dense than asthenosphere for a few tens of millions of years but after this becomes increasingly denser than asthenosphere. This is because the chemically differentiated oceanic crust is lighter than asthenosphere, but thermal contraction of the mantle lithosphere makes it more dense than the asthenosphere. The gravitational instability of mature oceanic lithosphere has the effect that at subduction zones, oceanic lithosphere invariably sinks underneath the overriding lithosphere, which can be oceanic or continental. New oceanic lithosphere is constantly being produced at mid-ocean ridges and is recycled back to the mantle at subduction zones. As a result, oceanic lithosphere is much younger than continental lithosphere: the oldest oceanic lithosphere is about 170 million years old, while parts of the continental lithosphere are billions of years old. The oldest parts of continental lithosphere underlie cratons, and the mantle lithosphere there is thicker and less dense than typical; the relatively low density of such mantle "roots of cratons" helps to stabilize these regions.

Subducted Lithosphere

Geophysical studies in the early 21st century posit that large pieces of the lithosphere have been subducted into the mantle as deep as 2900 km to near the core-mantle boundary, while others "float" in the upper mantle, while some stick down into the mantle as far as 400 km but remain "attached" to the continental plate above, similar to the extent of the "tectosphere" proposed by Jordan in 1988.

Mantle Xenoliths

Geoscientists can directly study the nature of the subcontinental mantle by examining mantle xenoliths brought up in kimberlite, lamproite, and other volcanic pipes. The histories of these xenoliths have been investigated by many methods, including analyses of abundances of isotopes of osmium and rhenium. Such studies have confirmed that mantle lithospheres below some cratons have persisted for periods in excess of 3 billion years, despite the mantle flow that accompanies plate tectonics.

Cryosphere

Overview of the Cryosphere and its larger components, from the UN Environment Programme Global Outlook for Ice and Snow.

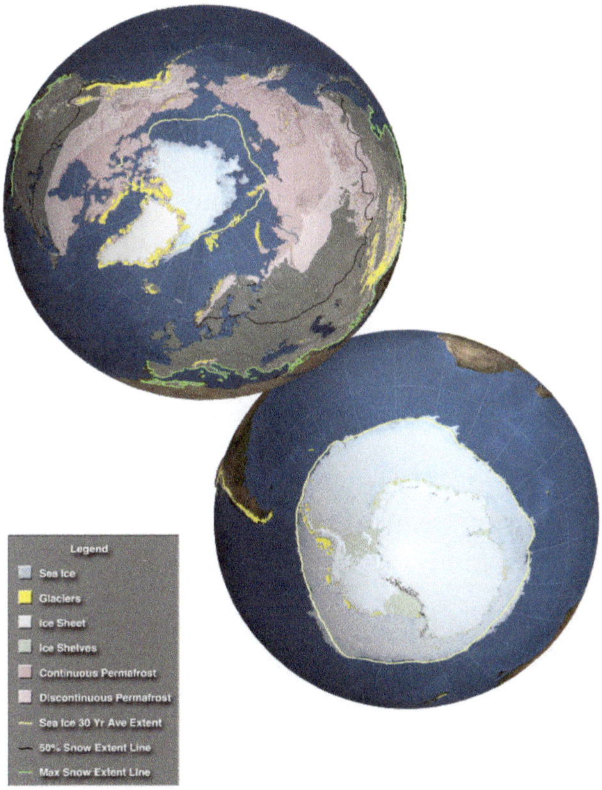

This high resolution image, designed for the Fifth Assessment Report of the IPCC, shows the extent of the regions affected by components of the cryosphere around the world. Over land, continuous permafrost is shown in a dark pink while discontinuous permafrost is shown in a lighter shade of pink. Over much of the northern hemisphere's land area, a semi-transparent white veil depicts the regions that are affected by snowfall at least one day during the period 2000-2012. The bright green line along the southern border of this region shows the maximum snow extent while a black line across the North America, Europe and Asia shows the 50% snow extent line. Glaciers are shown as small golden dots in mountainous areas and in the far northern and southern latitudes. Over the water, ice shelves are shown around Antarctica along with sea ice surrounding the ice shelves. Sea ice is also shown at the North Pole. For both poles. the 30 year average sea ice extent is shown by a yellow outline. In addition, the ice sheets of Greenland and Antarctica are clearly visible.

The cryosphere is those portions of Earth's surface where water is in solid form, including sea ice, lake ice, river ice, snow cover, glaciers, ice caps, ice sheets, and frozen ground (which includes permafrost). Thus, there is a wide overlap with the hydrosphere. The cryosphere is an integral part of the global climate system with important linkages and feedbacks generated through its influence on surface energy and moisture fluxes, clouds, precipitation, hydrology, atmospheric and oceanic circulation. Through these feedback processes, the cryosphere plays a significant role in the global climate and in climate model response to global changes. The term deglaciation describes the retreat of cryospheric features. Cryology is the study of cryospheres.

Structure

Frozen water is found on the Earth's surface primarily as snow cover, freshwater ice in lakes and rivers, sea ice, glaciers, ice sheets, and frozen ground and permafrost (permanently frozen ground). The residence time of water in each of these cryospheric sub-systems varies widely. Snow cover and freshwater ice are essentially seasonal, and most sea ice, except for ice in the central Arctic, lasts only a few years if it is not seasonal. A given water particle in glaciers, ice sheets, or ground ice, however, may remain frozen for 10-100,000 years or longer, and deep ice in parts of East Antarctica may have an age approaching 1 million years.

Most of the world's ice volume is in Antarctica, principally in the East Antarctic Ice Sheet. In terms of areal extent, however, Northern Hemisphere winter snow and ice extent comprise the largest area, amounting to an average 23% of hemispheric surface area in January. The large areal extent and the important climatic roles of snow and ice, related to their unique physical properties, indicate that the ability to observe and model snow and ice-cover extent, thickness, and physical properties (radiative and thermal properties) is of particular significance for climate research.

There are several fundamental physical properties of snow and ice that modulate energy exchanges between the surface and the atmosphere. The most important properties are the surface reflectance (albedo), the ability to transfer heat (thermal diffusivity), and the ability to change state (latent heat). These physical properties, together with surface roughness, emissivity, and dielectric characteristics, have important implications for observing snow and ice from space. For example, surface roughness is often the dominant factor determining the strength of radar backscatter . Physical properties such as crystal structure, density, length, and liquid water content are important factors affecting the transfers of heat and water and the scattering of microwave energy.

The surface reflectance of incoming solar radiation is important for the surface energy balance (SEB). It is the ratio of reflected to incident solar radiation, commonly referred to as albedo. Climatologists are primarily interested in albedo integrated over the shortwave portion of the electromagnetic spectrum (~300 to 3500 nm), which coincides with the main solar energy input. Typically, albedo values for non-melting snow-covered surfaces are high (~80-90%) except in the case of forests. The higher albedos for snow and ice cause rapid shifts in surface reflectivity in autumn and spring in high latitudes, but the overall climatic significance of this increase is spatially and temporally modulated by cloud cover. (Planetary albedo is determined principally by cloud cover, and by the small amount of total solar radiation received in high latitudes during winter months.) Summer and autumn are times of high-average cloudiness over the Arctic Ocean so the albedo feedback associated with the large seasonal changes in sea-ice extent is greatly reduced. Groisman *et al.* (1994a) observed that snow cover exhibited the greatest influence on the Earth radiative balance in the spring (April to May) period when incoming solar radiation was greatest over snow-covered areas.

The thermal properties of cryospheric elements also have important climatic consequences. Snow and ice have much lower thermal diffusivities than air. Thermal diffusivity is a measure of the speed at which temperature waves can penetrate a substance. Snow and ice are many orders of magnitude less efficient at diffusing heat than air. Snow cover insulates the ground surface, and sea ice insulates the underlying ocean, decoupling the surface-atmosphere interface with respect to both heat and moisture fluxes. The flux of moisture from a water surface is eliminated by even a thin skin of ice, whereas the flux of heat through thin ice continues to be substantial until it attains a thickness in excess of 30 to 40 cm. However, even a small amount of snow on top of the ice will dramatically reduce the heat flux and slow down the rate of ice growth. The insulating effect of snow also has major implications for the hydrological cycle. In non-permafrost regions, the insulating effect of snow is such that only near-surface ground freezes and deep-water drainage is uninterrupted.

While snow and ice act to insulate the surface from large energy losses in winter, they also act to retard warming in the spring and summer because of the large amount of energy required to melt ice (the latent heat of fusion, 3.34×10^5 J/kg at 0 °C). However, the strong static stability of the atmosphere over areas of extensive snow or ice tends to confine the immediate cooling effect to a relatively shallow layer, so that associated atmospheric anomalies are usually short-lived and local to regional in scale. In some areas of the world such as Eurasia, however, the cooling associated with a heavy snowpack and moist spring soils is known to play a role in modulating the summer monsoon circulation. Gutzler and Preston (1997) recently presented evidence for a similar snow-summer circulation feedback over the southwestern United States.

The role of snow cover in modulating the monsoon is just one example of a short-term cryosphere-climate feedback involving the land surface and the atmosphere. From Figure 1 it can be seen that there are numerous cryosphere-climate feedbacks in the global climate system. These operate over a wide range of spatial and temporal scales from local seasonal cooling of air temperatures to hemispheric-scale variations in ice sheets over time-scales of thousands of years. The feedback mechanisms involved are often complex and incompletely understood. For example, Curry *et al.* (1995) showed that the so-called "simple" sea ice-albedo feedback involved complex interactions with lead fraction, melt ponds, ice thickness, snow cover, and sea-ice extent.

Snow

Snow cover has the second-largest areal extent of any component of the cryosphere, with a mean maximum areal extent of approximately 47 million km². Most of the Earth's snow-covered area (SCA) is located in the Northern Hemisphere, and temporal variability is dominated by the seasonal cycle; Northern Hemisphere snow-cover extent ranges from 46.5 million km² in January to 3.8 million km² in August. North American winter SCA has exhibited an increasing trend over much of this century (Brown and Goodison 1996; Hughes *et al.* 1996) largely in response to an increase in precipitation. However, the available satellite data show that the hemispheric winter snow cover has exhibited little interannual variability over the 1972-1996 period, with a coefficient of variation (COV=s.d./mean) for January Northern Hemisphere snow cover of < 0.04. According to Groisman *et al.* (1994a) Northern Hemisphere spring snow cover should exhibit a decreasing trend to explain an observed increase in Northern Hemisphere spring air temperatures this century. Preliminary estimates of SCA from historical and reconstructed in situ snow-cover data suggest this is the case for Eurasia, but not for North America, where spring snow cover has remained close to current levels over most of this century. Because of the close relationship observed between hemispheric air temperature and snow-cover extent over the period of satellite data (IPCC 1996), there is considerable interest in monitoring Northern Hemisphere snow-cover extent for detecting and monitoring climate change.

Snow cover is an extremely important storage component in the water balance, especially seasonal snowpacks in mountainous areas of the world. Though limited in extent, seasonal snowpacks in the Earth's mountain ranges account for the major source of the runoff for stream flow and groundwater recharge over wide areas of the midlatitudes. For example, over 85% of the annual runoff from the Colorado River basin originates as snowmelt. Snowmelt runoff from the Earth's mountains fills the rivers and recharges the aquifers that over a billion people depend on for their water resources. Further, over 40% of the world's protected areas are in mountains, attesting to their value both as unique ecosystems needing protection and as recreation areas for humans. Climate warming is expected to result in major changes to the partitioning of snow and rainfall, and to the timing of snowmelt, which will have important implications for water use and management. These changes also involve potentially important decadal and longer time-scale feedbacks to the climate system through temporal and spatial changes in soil moisture and runoff to the oceans.(Walsh 1995). Freshwater fluxes from the snow cover into the marine environment may be important, as the total flux is probably of the same magnitude as desalinated ridging and rubble areas of sea ice. In addition, there is an associated pulse of precipitated pollutants which accumulate over the Arctic winter in snowfall and are released into the ocean upon ablation of the sea-ice .

Sea Ice

Sea ice covers much of the polar oceans and forms by freezing of sea water. Satellite data since the early 1970s reveal considerable seasonal, regional, and interannual variability in the sea-ice covers of both hemispheres. Seasonally, sea-ice extent in the Southern Hemisphere varies by a factor of 5, from a minimum of 3-4 million km^2 in February to a maximum of 17-20 million km^2 in September. The seasonal variation is much less in the Northern Hemisphere where the confined nature and high latitudes of the Arctic Ocean result in a much larger perennial ice cover, and the surrounding land limits the equatorward extent of wintertime ice. Thus, the seasonal variability in Northern Hemisphere ice extent varies by only a factor of 2, from a minimum of 7-9 million km^2 in September to a maximum of 14-16 million km^2 in March.

The ice cover exhibits much greater regional-scale interannual variability than it does hemispherical. For instance, in the region of the Sea of Okhotsk and Japan, maximum ice extent decreased from 1.3 million km^2 in 1983 to 0.85 million km^2 in 1984, a decrease of 35%, before rebounding the following year to 1.2 million km^2. The regional fluctuations in both hemispheres are such that for any several-year period of the satellite record some regions exhibit decreasing ice coverage while others exhibit increasing ice cover. The overall trend indicated in the passive microwave record from 1978 through mid-1995 shows that the extent of Arctic sea ice is decreasing 2.7% per decade. Subsequent work with the satellite passive-microwave data indicates that from late October 1978 through the end of 1996 the extent of Arctic sea ice decreased by 2.9% per decade while the extent of Antarctic sea ice increased by 1.3% per decade. The Intergovernmental Panel on Climate Change publication *Climate change 2013: The Physical Science Basis* stated that sea ice extent for the Northern Hemisphere showed a decrease of 3.8% ± 0.3% per decade from November 1978 to December 2012.

Lake Ice and River Ice

Ice forms on rivers and lakes in response to seasonal cooling. The sizes of the ice bodies involved are too small to exert other than localized climatic effects. However, the freeze-up/break-up processes respond to large-scale and local weather factors, such that considerable interannual variability exists in the dates of appearance and disappearance of the ice. Long series of lake-ice observations can serve as a proxy climate record, and the monitoring of freeze-up and break-up trends may provide a convenient integrated and seasonally specific index of climatic perturbations. Information on river-ice conditions is less useful as a climatic proxy because ice formation is strongly dependent on river-flow regime, which is affected by precipitation, snow melt, and watershed runoff as well as being subject to human interference that directly modifies channel flow, or that indirectly affects the runoff via land-use practices.

Lake freeze-up depends on the heat storage in the lake and therefore on its depth, the rate and temperature of any inflow, and water-air energy fluxes. Information on lake depth is often unavailable, although some indication of the depth of shallow lakes in the Arctic can be obtained from airborne radar imagery during late winter (Sellman *et al.* 1975) and space-borne optical imagery during summer (Duguay and Lafleur 1997). The timing of breakup is modified by snow depth on the ice as well as by ice thickness and freshwater inflow.

Frozen Ground and Permafrost

Frozen ground (permafrost and seasonally frozen ground) occupies approximately 54 million km^2 of the exposed land areas of the Northern Hemisphere (Zhang et al., 2003) and therefore has the largest areal extent of any component of the cryosphere. Permafrost (perennially frozen ground) may occur where mean annual air temperatures (MAAT) are less than -1 or -2 °C and is generally continuous where MAAT are less than -7 °C. In addition, its extent and thickness are affected by ground moisture content, vegetation cover, winter snow depth, and aspect. The global extent of permafrost is still not completely known, but it underlies approximately 20% of Northern Hemisphere land areas. Thicknesses exceed 600 m along the Arctic coast of northeastern Siberia and Alaska, but, toward the margins, permafrost becomes thinner and horizontally discontinuous. The marginal zones will be more immediately subject to any melting caused by a warming trend. Most of the presently existing permafrost formed during previous colder conditions and is therefore relic. However, permafrost may form under present-day polar climates where glaciers retreat or land emergence exposes unfrozen ground. Washburn (1973) concluded that most continuous permafrost is in balance with the present climate at its upper surface, but changes at the base depend on the present climate and geothermal heat flow; in contrast, most discontinuous permafrost is probably unstable or "in such delicate equilibrium that the slightest climatic or surface change will have drastic disequilibrium effects".

Under warming conditions, the increasing depth of the summer active layer has significant impacts on the hydrologic and geomorphic regimes. Thawing and retreat of permafrost have been reported in the upper Mackenzie Valley and along the southern margin of its occurrence in Manitoba, but such observations are not readily quantified and generalized. Based on average latitudinal gradients of air temperature, an average northward displacement of the southern permafrost boundary by 50-to-150 km could be expected, under equilibrium conditions, for a 1 °C warming.

Only a fraction of the permafrost zone consists of actual ground ice. The remainder (dry permafrost) is simply soil or rock at subfreezing temperatures. The ice volume is generally greatest in the uppermost permafrost layers and mainly comprises pore and segregated ice in Earth material. Measurements of bore-hole temperatures in permafrost can be used as indicators of net changes in temperature regime. Gold and Lachenbruch (1973) infer a 2-4 °C warming over 75 to 100 years at Cape Thompson, Alaska, where the upper 25% of the 400-m thick permafrost is unstable with respect to an equilibrium profile of temperature with depth (for the present mean annual surface temperature of -5 °C). Maritime influences may have biased this estimate, however. At Prudhoe Bay similar data imply a 1.8 °C warming over the last 100 years (Lachenbruch *et al.* 1982). Further complications may be introduced by changes in snow-cover depths and the natural or artificial disturbance of the surface vegetation.

The potential rates of permafrost thawing have been established by Osterkamp (1984) to be two centuries or less for 25-meter-thick permafrost in the discontinuous zone of interior Alaska, assuming warming from -0.4 to 0 °C in 3–4 years, followed by a further 2.6 °C rise. Although the response of permafrost (depth) to temperature change is typically a very slow process (Osterkamp 1984; Koster 1993), there is ample evidence for the fact that the active layer thickness quickly responds to a temperature change (Kane *et al.* 1991). Whether, under a warming or cooling scenario, global climate change will have a significant effect on the duration of frost-free periods in both regions with seasonally and perennially frozen ground.

Glaciers and Ice Sheets

Ice sheets and glaciers are flowing ice masses that rest on solid land. They are controlled by snow accumulation, surface and basal melt, calving into surrounding oceans or lakes and internal dynamics. The latter results from gravity-driven creep flow ("glacial flow") within the ice body and sliding on the underlying land, which leads to thinning and horizontal spreading. Any imbalance of this dynamic equilibrium between mass gain, loss and transport due to flow results in either growing or shrinking ice bodies.

Ice sheets are the greatest potential source of global freshwater, holding approximately 77% of the global total. This corresponds to 80 m of world sea-level equivalent, with Antarctica accounting for 90% of this. Greenland accounts for most of the remaining 10%, with other ice bodies and glaciers accounting for less than 0.5%. Because of their size in relation to annual rates of snow accumulation and melt, the residence time of water in ice sheets can extend to 100,000 or 1 million years. Consequently, any climatic perturbations produce slow responses, occurring over glacial and interglacial periods. Valley glaciers respond rapidly to climatic fluctuations with typical response times of 10–50 years. However, the response of individual glaciers may be asynchronous to the same climatic forcing because of differences in glacier length, elevation, slope, and speed of motion. Oerlemans (1994) provided evidence of coherent global glacier retreat which could be explained by a linear warming trend of 0.66 °C per 100 years.

While glacier variations are likely to have minimal effects upon global climate, their recession may have contributed one third to one half of the observed 20th Century rise in sea level (Meier 1984; IPCC 1996). Furthermore, it is extremely likely that such extensive glacier recession as is currently observed in the Western Cordillera of North America, where runoff from glacierized basins is used for irrigation and hydropower, involves significant hydrological and ecosystem impacts. Effective water-resource planning and impact mitigation in such areas depends upon developing a sophisticated knowledge of the status of glacier ice and the mechanisms that cause it to change. Furthermore, a clear understanding of the mechanisms at work is crucial to interpreting the global-change signals that are contained in the time series of glacier mass balance records.

Combined glacier mass balance estimates of the large ice sheets carry an uncertainty of about 20%. Studies based on estimated snowfall and mass output tend to indicate that the ice sheets are near balance or taking some water out of the oceans. Marinebased studies suggest sea-level rise from the Antarctic or rapid ice-shelf basal melting. Some authors (Paterson 1993; Alley 1997) have suggested that the difference between the observed rate of sea-level rise (roughly 2 mm/y) and the explained rate of sea-level rise from melting of mountain glaciers, thermal expansion of the ocean, etc. (roughly 1 mm/y or less) is similar to the modeled imbalance in the Antarctic (roughly 1 mm/y of sea-level rise; Huybrechts 1990), suggesting a contribution of sea-level rise from the Antarctic.

Relationships between global climate and changes in ice extent are complex. The mass balance of land-based glaciers and ice sheets is determined by the accumulation of snow, mostly in winter, and warm-season ablation due primarily to net radiation and turbulent heat fluxes to melting ice and snow from warm-air advection,(Munro 1990). However, most of Antarctica never experiences surface melting. Where ice masses terminate in the ocean, iceberg calving is the major contributor to mass loss. In this situation, the ice margin may extend out into deep water as a floating ice shelf, such as that in the Ross Sea. Despite the possibility that global warming could result in losses to the Greenland ice sheet being offset by gains to the Antarctic ice sheet, there is major concern about the possibility of a West Antarctic Ice Sheet collapse. The West Antarctic Ice Sheet is grounded on bedrock below sea level, and its collapse has the potential of raising the world sea level 6–7 m over a few hundred years.

Most of the discharge of the West Antarctic Ice Sheet is via the five major ice streams (faster flowing ice) entering the Ross Ice Shelf, the Rutford Ice Stream entering Ronne-Filchner shelf of the Weddell Sea, and the Thwaites Glacier and Pine Island Glacier entering the Amundsen Ice Shelf. Opinions differ as to the present mass balance of these systems (Bentley 1983, 1985), principally because of the limited data. The West Antarctic Ice Sheet is stable so long as the Ross Ice Shelf is constrained by drag along its lateral boundaries and pinned by local grounding.

References

- Kennish, Michael J. (2001). Practical handbook of marine science. Marine science series (3rd ed.). CRC Press. p. 35. ISBN 0-8493-2391-6.
- Wallace, John M. and Peter V. Hobbs. Atmospheric Science; An Introductory Survey.Elsevier. Second Edition, 2006. ISBN 978-0-12-732951-2.
- Lutgens, Frederick K. and Edward J. Tarbuck (1995) The Atmosphere, Prentice Hall, 6th ed., pp14-17, ISBN 0-13-350612-6

- Skinner, B.J. & Porter, S.C.: Physical Geology, page 17, chapt. The Earth: Inside and Out, 1987, John Wiley & Sons, ISBN 0-471-05668-5
- Greve, R.; Blatter, H. (2009). Dynamics of Ice Sheets and Glaciers. Springer. doi:10.1007/978-3-642-03415-2. ISBN 978-3-642-03414-5.
- "Climate Change 2013: The Physical Science Basis" (PDF). ipcc. Intergovernmental Panel on Climate Change. p. 324. Retrieved 16 June 2015.
- "Atmospheric Temperature Trends, 1979-2005 : Image of the Day". Earthobservatory.nasa.gov. 2000-01-01. Retrieved 2014-06-10.
- Pasyanos M. E. (2008-05-15). "Lithospheric Thickness Modeled from Long Period Surface Wave Dispersion" (PDF). Retrieved 2014-04-25.
- World Water Resources: A New Appraisal and Assessment for the 21st Century (Report). UNESCO. 1998. Retrieved 13 June 2013.
- Zimmer, Carl (3 October 2013). "Earth's Oxygen: A Mystery Easy to Take for Granted". New York Times. Retrieved 3 October 2013.
- ""homosphere"—AMS Glossary". Amsglossary.allenpress.com. Archived from the original on 14 September 2010. Retrieved 2010-10-16.

3

Ecosystem: An Overview

A community of living organisms is known as an ecosystem. The components that make up life on Earth can be divided into biotic and abiotic components; these are linked together through energy flows and natural cycles. This section is an overview of the subject matter incorporating all the major aspects of ecosystem.

Ecosystem

An ecosystem is a community of living organisms in conjunction with the nonliving components of their environment (things like air, water and mineral soil), interacting as a system. These biotic and abiotic components are regarded as linked together through nutrient cycles and energy flows. As ecosystems are defined by the network of interactions among organisms, and between organisms and their environment, they can be of any size but usually encompass specific, limited spaces (although some scientists say that the entire planet is an ecosystem).

Coral reefs are a highly productive marine ecosystem.

Rainforest ecosystems are rich in biodiversity. This is the Gambia River in Senegal's Niokolo-Koba National Park.

Energy, water, nitrogen and soil minerals are other essential abiotic components of an ecosystem. The energy that flows through ecosystems is obtained primarily from the sun. It generally enters the system through photosynthesis, a process that also captures carbon from the atmosphere. By feeding on plants and on one another, animals play an important role in the movement of matter and energy through the system. They also influence the quantity of plant and microbial biomass present. By breaking down dead organic matter, decomposers release carbon back to the atmosphere and facilitate nutrient cycling by converting nutrients stored in dead biomass back to a form that can be readily used by plants and other microbes.

Ecosystems are controlled both by external and internal factors. External factors such as climate, the parent material that forms the soil, and topography control the overall structure of an ecosystem and the way things work within it, but are not themselves influenced by the ecosystem. Other external factors include time and potential biota. Ecosystems are dynamic entities—invariably, they are subject to periodic disturbances and are in the process of recovering from some past disturbance. Ecosystems in similar environments that are located in different parts of the world can have very different characteristics simply because they contain different species. The introduction of non-native species can cause substantial shifts in ecosystem function. Internal factors not only control ecosystem processes but are also controlled by them and are often subject to feedback loops. While the resource inputs are generally controlled by external processes like climate and parent material, the availability of these resources within the ecosystem is controlled by internal factors like decomposition, root competition or shading. Other internal factors include disturbance, succession and the types of species present. Although humans exist and operate within ecosystems, their cumulative effects are large enough to influence external factors like climate.

Biodiversity affects ecosystem function, as do the processes of disturbance and succession. Ecosystems provide a variety of goods and services upon which people depend; the principles of ecosystem management suggest that rather than managing individual species, natural resources should be managed at the level of the ecosystem itself. Classifying ecosystems into ecologically homogeneous units is an important step towards effective ecosystem management, but there is no single, agreed-upon way to do this.

History and Development

The term "ecosystem" was first used in a publication by British ecologist Arthur Tansley. Tansley devised the concept to draw attention to the importance of transfers of materials between organisms and their environment. He later refined the term, describing it as "The whole system, ... including not only the organism-complex, but also the whole complex of physical factors forming what we call the environment". Tansley regarded ecosystems not simply as natural units, but as mental isolates. Tansley later defined the spatial extent of ecosystems using the term ecotope.

G. Evelyn Hutchinson, a pioneering limnologist who was a contemporary of Tansley's, combined Charles Elton's ideas about trophic ecology with those of Russian geochemist Vladimir Vernadsky to suggest that mineral nutrient availability in a lake limited algal production which would, in turn, limit the abundance of animals that feed on algae. Raymond Lindeman took these ideas one step further to suggest that the flow of energy through a lake was the primary driver of the ecosystem. Hutchinson's students, brothers Howard T. Odum and Eugene P. Odum, further developed a "systems approach" to the study of ecosystems, allowing them to study the flow of energy and material through ecological systems.

Ecosystem Processes

Energy and carbon enter ecosystems through photosynthesis, are incorporated into living tissue, transferred to other organisms that feed on the living and dead plant matter, and eventually released through respiration. Most mineral nutrients, on the other hand, are recycled within ecosystems.

Ecosystems are controlled both by external and internal factors. External factors, also called state factors, control the overall structure of an ecosystem and the way things work within it, but are not themselves influenced by the ecosystem. The most important of these is climate. Climate determines the biome in which the ecosystem is embedded. Rainfall patterns and temperature seasonality determine the amount of water available to the ecosystem and the supply of energy available (by influencing photosynthesis). Parent material, the underlying geological material that gives rise to soils, determines the nature of the soils present, and influences the supply of mineral nutrients. Topography also controls ecosystem processes by affecting things like microclimate, soil development and the movement of water through a system. This may be the difference between the ecosystem present in wetland situated in a small depression on the landscape, and one present on an adjacent steep hillside.

Other external factors that play an important role in ecosystem functioning include time and potential biota. Ecosystems are dynamic entities—invariably, they are subject to periodic disturbances and are in the process of recovering from some past disturbance. Time plays a role in the development of soil from bare rock and the recovery of a community from disturbance. Similarly, the set of organisms that can potentially be present in an area can also have a major impact on ecosystems. Ecosystems in similar environments that are located in different parts of the world can end up doing things very differently simply because they have different pools of species present. The introduction of non-native species can cause substantial shifts in ecosystem function.

Unlike external factors, internal factors in ecosystems not only control ecosystem processes, but are also controlled by them. Consequently, they are often subject to feedback loops. While the resource inputs are generally controlled by external processes like climate and parent material, the availability of these resources within the ecosystem is controlled by internal factors like decomposition, root competition or shading. Other factors like disturbance, succession or the types of species present are also internal factors. Human activities are important in almost all ecosystems. Although humans exist and operate within ecosystems, their cumulative effects are large enough to influence external factors like climate.

Primary Production

Primary production is the production of organic matter from inorganic carbon sources. Overwhelmingly, this occurs through photosynthesis. The energy incorporated through this process supports life on earth, while the carbon makes up much of the organic matter in living and dead biomass, soil carbon and fossil fuels. It also drives the carbon cycle, which influences global climate via the greenhouse effect.

Through the process of photosynthesis, plants capture energy from light and use it to combine carbon dioxide and water to produce carbohydrates and oxygen. The photosynthesis carried out by all the plants in an ecosystem is called the gross primary production (GPP). About 48–60% of the GPP is consumed in plant respiration. The remainder, that portion of GPP that is not used up by respiration, is known as the net primary production (NPP). Total photosynthesis is limited by a range of environmental factors. These include the amount of light available, the amount of leaf area a plant has to capture light (shading by other plants is a major limitation of photosynthesis), rate at which carbon dioxide can be supplied to the chloroplasts to support photosynthesis, the availability of water, and the availability of suitable temperatures for carrying out photosynthesis.

Energy Flow

The carbon and energy incorporated into plant tissues (net primary production) is either consumed by animals while the plant is alive, or it remains uneaten when the plant tissue dies and becomes detritus. In terrestrial ecosystems, roughly 90% of the NPP ends up being broken down by decomposers. The remainder is either consumed by animals while still alive and enters the plant-based trophic system, or it is consumed after it has died, and enters the detritus-based trophic system. In aquatic systems, the proportion of plant biomass that gets consumed by herbivores is much higher. In trophic systems photosynthetic organisms are the primary producers. The organisms that consume their tissues are called primary consumers or secondary producers—herbivores. Organisms which feed on microbes (bacteria and fungi) are termed microbivores. Animals that feed on primary consumers—carnivores—are secondary consumers. Each of these constitutes a trophic level. The sequence of consumption—from plant to herbivore, to carnivore—forms a food chain. Real systems are much more complex than this—organisms will generally feed on more than one form of food, and may feed at more than one trophic level. Carnivores may capture some prey which are part of a plant-based trophic system and others that are part of a detritus-based trophic system (a bird that feeds both on herbivorous grasshoppers and earthworms, which consume detritus). Real systems, with all these complexities, form food webs rather than food chains.

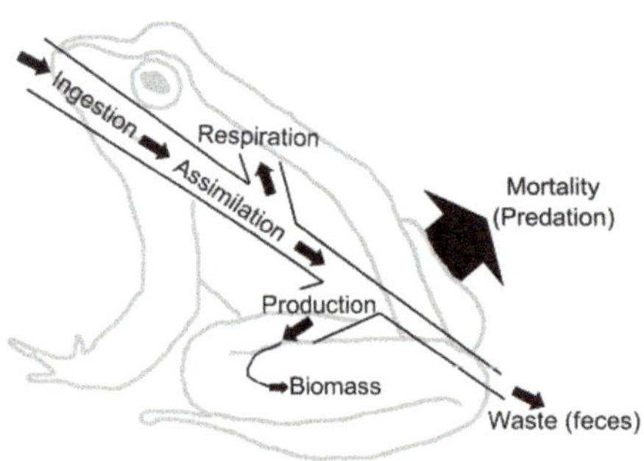

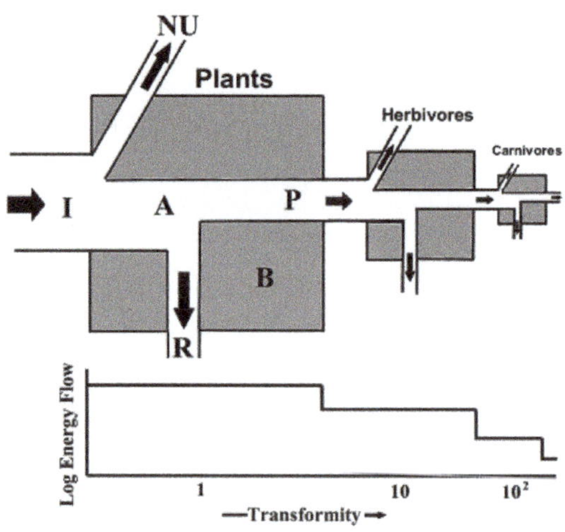

Left: Energy flow diagram of a frog. The frog represents a node in an extended food web. The energy ingested is utilized for metabolic processes and transformed into biomass. The energy flow continues on its path if the frog is ingested by predators, parasites, or as a decaying carcass in soil. This energy flow diagram illustrates how energy is lost as it fuels the metabolic process that transforms the energy and nutrients into biomass.

Right: An expanded three link energy food chain (1. plants, 2. herbivores, 3. carnivores) illustrating the relationship between food flow diagrams and energy transformity. The transformity of energy becomes degraded, dispersed, and diminished from higher quality to lesser quantity as the energy within a food chain flows from one trophic species into another. Abbreviations: I=input, A=assimilation, R=respiration, NU=not utilized, P=production, B=biomass.

Decomposition

The carbon and nutrients in dead organic matter are broken down by a group of processes known as decomposition. This releases nutrients that can then be re-used for plant and microbial production, and returns carbon dioxide to the atmosphere (or water) where it can be used for photosynthesis. In the absence of decomposition, dead organic matter would accumulate in an ecosystem and nutrients and atmospheric carbon dioxide would be depleted. Approximately 90% of terrestrial NPP goes directly from plant to decomposer.

Decomposition processes can be separated into three categories—leaching, fragmentation and chemical alteration of dead material. As water moves through dead organic matter, it dissolves and carries with it the water-soluble components. These are then taken up by organisms in the soil, react with mineral soil, or are transported beyond the confines of the ecosystem (and are considered "lost" to it). Newly shed leaves and newly dead animals have high concentrations of water-soluble components, and include sugars, amino acids and mineral nutrients. Leaching is more important in wet environments, and much less important in dry ones.

Fragmentation processes break organic material into smaller pieces, exposing new surfaces for colonization by microbes. Freshly shed leaf litter may be inaccessible due to an outer layer of cuticle or bark, and cell contents are protected by a cell wall. Newly dead animals may be covered by an exoskeleton. Fragmentation processes, which break through these protective layers, accelerate the rate of microbial decomposition. Animals fragment detritus as they hunt for food, as does passage through the gut. Freeze-thaw cycles and cycles of wetting and drying also fragment dead material.

The chemical alteration of dead organic matter is primarily achieved through bacterial and fungal action. Fungal hyphae produce enzymes which can break through the tough outer structures surrounding dead plant material. They also produce enzymes which break down lignin, which allows to them access to both cell contents and to the nitrogen in the lignin. Fungi can transfer carbon and nitrogen through their hyphal networks and thus, unlike bacteria, are not dependent solely on locally available resources.

Decomposition rates vary among ecosystems. The rate of decomposition is governed by three sets of factors—the physical environment (temperature, moisture and soil properties), the quantity and quality of the dead material available to decomposers, and the nature of the microbial community itself. Temperature controls the rate of microbial respiration; the higher the temperature, the faster microbial decomposition occurs. It also affects soil moisture, which slows microbial growth and reduces leaching. Freeze-thaw cycles also affect decomposition—freezing temperatures kill soil microorganisms, which allows leaching to play a more important role in moving nutrients around. This can be especially important as the soil thaws in the Spring, creating a pulse of nutrients which become available.

Decomposition rates are low under very wet or very dry conditions. Decomposition rates are highest in wet, moist conditions with adequate levels of oxygen. Wet soils tend to become deficient in oxygen (this is especially true in wetlands), which slows microbial growth. In dry soils, decomposition slows as well, but bacteria continue to grow (albeit at a slower rate) even after soils become too dry to support plant growth. When the rains return and soils become wet, the osmotic gradient between the bacterial cells and the soil water causes the cells to gain water quickly. Under these conditions, many bacterial cells burst, releasing a pulse of nutrients. Decomposition rates also tend to be slower in acidic soils. Soils which are rich in clay minerals tend to have lower decomposition rates, and thus, higher levels of organic matter. The smaller particles of clay result in a larger surface area that can hold water. The higher the water content of a soil, the lower the oxygen content and consequently, the lower the rate of decomposition. Clay minerals also bind particles of organic material to their surface, making them less accessibly to microbes. Soil disturbance like tilling increase decomposition by increasing the amount of oxygen in the soil and by exposing new organic matter to soil microbes.

The quality and quantity of the material available to decomposers is another major factor that influences the rate of decomposition. Substances like sugars and amino acids decompose readily and are considered "labile". Cellulose and hemicellulose, which are broken down more slowly, are "moderately labile". Compounds which are more resistant to decay, like lignin or cutin, are considered "recalcitrant". Litter with a higher proportion of labile compounds decomposes much more rapidly than does litter with a higher proportion of recalcitrant material. Consequently, dead animals decompose more rapidly than dead leaves, which themselves decompose more rapidly than fallen branches. As organic material in the soil ages, its quality decreases. The more labile compounds decompose quickly, leaving an increasing proportion of recalcitrant material. Microbial cell walls also contain recalcitrant materials like chitin, and these also accumulate as the microbes die, further reducing the quality of older soil organic matter.

Nutrient Cycling

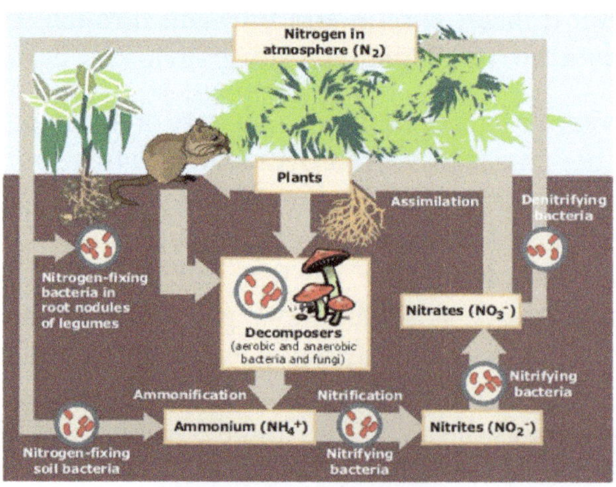

Biological nitrogen cycling

Ecosystems continually exchange energy and carbon with the wider environment; mineral nutrients, on the other hand, are mostly cycled back and forth between plants, animals, microbes and the soil. Most nitrogen enters ecosystems through biological nitrogen fixation, is deposited through precipitation, dust, gases or is applied as fertilizer. Since most terrestrial ecosystems are nitrogen-limited, nitrogen cycling is an important control on ecosystem production.

Until modern times, nitrogen fixation was the major source of nitrogen for ecosystems. Nitrogen fixing bacteria either live symbiotically with plants, or live freely in the soil. The energetic cost is high for plants which support nitrogen-fixing symbionts—as much as 25% of GPP when measured in controlled conditions. Many members of the legume plant family support nitrogen-fixing symbionts. Some cyanobacteria are also capable of nitrogen fixation. These are phototrophs, which carry out photosynthesis. Like other nitrogen-fixing bacteria, they can either be free-living or have symbiotic relationships with plants. Other sources of nitrogen include acid deposition produced through the combustion of fossil fuels, ammonia gas which evaporates from agricultural fields which have had fertilizers applied to them, and dust. Anthropogenic nitrogen inputs account for about 80% of all nitrogen fluxes in ecosystems.

When plant tissues are shed or are eaten, the nitrogen in those tissues becomes available to animals and microbes. Microbial decomposition releases nitrogen compounds from dead organic matter in the soil, where plants, fungi and bacteria compete for it. Some soil bacteria use organic nitrogen-containing compounds as a source of carbon, and release ammonium ions into the soil. This process is known as nitrogen mineralization. Others convert ammonium to nitrite and nitrate ions, a process known as nitrification. Nitric oxide and nitrous oxide are also produced during nitrification. Under nitrogen-rich and oxygen-poor conditions, nitrates and nitrites are converted to nitrogen gas, a process known as denitrification.

Other important nutrients include phosphorus, sulfur, calcium, potassium, magnesium and manganese. Phosphorus enters ecosystems through weathering. As ecosystems age this supply diminishes, making phosphorus-limitation more common in older landscapes (especially in the tropics). Calcium and sulfur are also produced by weathering, but acid deposition is an important source of sulfur in many ecosystems. Although magnesium and manganese are produced by weathering, exchanges between soil organic matter and living cells account for a significant portion of ecosystem fluxes. Potassium is primarily cycled between living cells and soil organic matter.

Function and Biodiversity

Ecosystem processes are broad generalizations that actually take place through the actions of individual organisms. The nature of the organisms—the species, functional groups and trophic levels to which they belong—dictates the sorts of actions these individuals are capable of carrying out, and the relative efficiency with which they do so. Thus, ecosystem processes are driven by the number of species in an ecosystem, the exact nature of each individual species, and the relative abundance organisms within these species. Biodiversity plays an important role in ecosystem functioning.

Loch Lomond in Scotland forms a relatively isolated ecosystem. The fish community of this lake has remained stable over a long period until a number of introductions in the 1970s restructured its food web.

Spiny forest at Ifaty, Madagascar, featuring various *Adansonia* (baobab) species, *Alluaudia procera* (Madagascar ocotillo) and other vegetation.

Ecological theory suggests that in order to coexist, species must have some level of limiting similarity—they must be different from one another in some fundamental way, otherwise one species would competitively exclude the other. Despite this, the cumulative effect of additional species in an ecosystem is not linear—additional species may enhance nitrogen retention, for example, but beyond some level of species richness, additional species may have little additive effect. The addition (or loss) of species which are ecologically similar to those already present in an ecosystem tends to only have a small effect on ecosystem function. Ecologically distinct species, on the other hand, have a much larger effect. Similarly, dominant species have a large impact on ecosystem function, while rare species tend to have a small effect. Keystone species tend to have an effect on ecosystem function that is disproportionate to their abundance in an ecosystem.

Ecosystem Goods and Services

Ecosystems provide a variety of goods and services upon which people depend. Ecosystem goods include the "tangible, material products" of ecosystem processes—food, construction material, medicinal plants—in addition to less tangible items like tourism and recreation, and genes from wild plants and animals that can be used to improve domestic species. Ecosystem services, on the other hand, are generally "improvements in the condition or location of things of value". These include things like the maintenance of hydrological cycles, cleaning air and water, the maintenance of oxygen in the atmosphere, crop pollination and even things like beauty, inspiration and opportunities for research. While ecosystem goods have traditionally been recognized as being the basis for things of economic value, ecosystem services tend to be taken for granted. While Gretchen Daily's original definition distinguished between ecosystem goods and ecosystem services, Robert Costanza and colleagues' later work and that of the Millennium Ecosystem Assessment lumped all of these together as ecosystem services.

Ecosystem Management

When natural resource management is applied to whole ecosystems, rather than single species, it is termed ecosystem management. A variety of definitions exist: F. Stuart Chapin and coauthors define it as "the application of ecological science to resource management to promote long-term sustainability of ecosystems and the delivery of essential ecosystem goods and services", while Norman Christensen and coauthors defined it as "management driven by explicit goals, executed by policies, protocols, and practices, and made adaptable by monitoring and research based on our best understanding of the ecological interactions and processes necessary to sustain ecosystem structure and function" and Peter Brussard and colleagues defined it as "managing areas at various scales in such a way that ecosystem services and biological resources are preserved while appropriate human use and options for livelihood are sustained".

Although definitions of ecosystem management abound, there is a common set of principles which underlie these definitions. A fundamental principle is the long-term sustainability of the production of goods and services by the ecosystem; "intergenerational sustainability [is] a precondition for management, not an afterthought". It also requires clear goals with respect to future trajectories and behaviors of the system being managed. Other important requirements include a sound ecological understanding of the system, including connectedness, ecological dynamics and the context in which the system is embedded. Other important principles include an understanding of the role of humans as components of the ecosystems and the use of adaptive management. While ecosystem management can be used as part of a plan for wilderness conservation, it can also be used in intensively managed ecosystems.

Ecosystem Dynamics

Ecosystems are dynamic entities—invariably, they are subject to periodic disturbances and are in the process of recovering from some past disturbance. When an ecosystem is subject to some sort of perturbation, it responds by moving away from its initial state. The tendency of a system to remain close to its equilibrium state, despite that disturbance, is termed its resistance. On the other hand, the speed with which it returns to its initial state after disturbance is called its resilience.

Temperate rainforest on the Olympic Peninsula in Washington state.

The High Peaks Wilderness Area in the 6,000,000-acre (2,400,000 ha) Adirondack Park is an example of a diverse ecosystem.

From one year to another, ecosystems experience variation in their biotic and abiotic environments. A drought, an especially cold winter and a pest outbreak all constitute short-term variability in environmental conditions. Animal populations vary from year to year, building up during resource-rich periods and crashing as they overshoot their food supply. These changes play out in changes in NPP, decomposition rates, and other ecosystem processes. Longer-term changes also shape ecosystem processes—the forests of eastern North America still show legacies of cultivation which ceased 200 years ago, while methane production in eastern Siberian lakes is controlled by organic matter which accumulated during the Pleistocene.

Disturbance also plays an important role in ecological processes. F. Stuart Chapin and coauthors define disturbance as "a relatively discrete event in time and space that alters the structure of populations, communities and ecosystems and causes changes in resources availability or the physical environment". This can range from tree falls and insect outbreaks to hurricanes and wildfires to volcanic eruptions and can cause large changes in plant, animal and microbe populations, as well soil organic matter content. Disturbance is followed by succession, a "directional change in ecosystem structure and functioning resulting from biotically driven changes in resources supply."

The frequency and severity of disturbance determines the way it impacts ecosystem function. Major disturbance like a volcanic eruption or glacial advance and retreat leave behind soils that lack plants, animals or organic matter. Ecosystems that experience disturbances that undergo primary succession. Less severe disturbance like forest fires, hurricanes or cultivation result in secondary succession. More severe disturbance and more frequent disturbance result in longer recovery times. Ecosystems recover more quickly from less severe disturbance events.

The early stages of primary succession are dominated by species with small propagules (seed and spores) which can be dispersed long distances. The early colonizers—often algae, cyanobacteria and lichens—stabilize the substrate. Nitrogen supplies are limited in new soils, and nitrogen-fixing species tend to play an important role early in primary succession. Unlike in primary succession, the species that dominate secondary succession, are usually present from the start of the process, often in the soil seed bank. In some systems the successional pathways are fairly consistent, and thus, are easy to predict. In others, there are many possible pathways—for example, the introduced nitrogen-fixing legume, *Myrica faya*, alter successional trajectories in Hawaiian forests.

The theoretical ecologist Robert Ulanowicz has used information theory tools to describe the structure of ecosystems, emphasizing mutual information (correlations) in studied systems. Drawing on this methodology and prior observations of complex ecosystems, Ulanowicz depicts approaches to determining the stress levels on ecosystems and predicting system reactions to defined types of alteration in their settings (such as increased or reduced energy flow, and eutrophication.

Ecosystem Ecology

A hydrothermal vent is an ecosystem on the ocean floor. (The scale bar is 1 m.)

Ecosystem ecology studies "the flow of energy and materials through organisms and the physical environment". It seeks to understand the processes which govern the stocks of material and energy in ecosystems, and the flow of matter and energy through them. The study of ecosystems can cover 10 orders of magnitude, from the surface layers of rocks to the surface of the planet.

There is no single definition of what constitutes an ecosystem. German ecologist Ernst-Detlef Schulze and coauthors defined an ecosystem as an area which is "uniform regarding the biological turnover, and contains all the fluxes above and below the ground area under consideration." They explicitly reject Gene Likens' use of entire river catchments as "too wide a demarcation" to be a single ecosystem, given the level of heterogeneity within such an area. Other authors have suggested that an ecosystem can encompass a much larger area, even the whole planet. Schulze and coauthors also rejected the idea that a single rotting log could be studied as an ecosystem because the size of the flows between the log and its surroundings are too large, relative to the proportion cycles within the log. Philosopher of science Mark Sagoff considers the failure to define "the kind of object it studies" to be an obstacle to the development of theory in ecosystem ecology.

Ecosystems can be studied through a variety of approaches—theoretical studies, studies monitoring specific ecosystems over long periods of time, those that look at differences between ecosystems to elucidate how they work and direct manipulative experimentation. Studies can be carried out at a variety of scales, from microcosms and mesocosms which serve as simplified representations of ecosystems, through whole-ecosystem studies. American ecologist Stephen R. Carpenter has argued that microcosm experiments can be "irrelevant and diversionary" if they are not carried out in conjunction with field studies carried out at the ecosystem scale, because microcosm experiments often fail to accurately predict ecosystem-level dynamics.

The Hubbard Brook Ecosystem Study, established in the White Mountains, New Hampshire in 1963, was the first successful attempt to study an entire watershed as an ecosystem. The study used stream chemistry as a means of monitoring ecosystem properties, and developed a detailed biogeochemical model of the ecosystem. Long-term research at the site led to the discovery of acid rain in North America in 1972, and was able to document the consequent depletion of soil cations (especially calcium) over the next several decades.

Classification

Classifying ecosystems into ecologically homogeneous units is an important step towards effective ecosystem management. A variety of systems exist, based on vegetation cover, remote sensing, and bioclimatic classification systems. American geographer Robert Bailey defines a hierarchy of ecosystem units ranging from microecosystems (individual homogeneous sites, on the order of 10 square kilometres (4 sq mi) in area), through mesoecosystems (landscape mosaics, on the order of 1,000 square kilometres

(400 sq mi)) to macroecosystems (ecoregions, on the order of 100,000 square kilometres (40,000 sq mi)).

Flora of Baja California Desert, Cataviña region, Mexico.

Bailey outlined five different methods for identifying ecosystems: *gestalt* ("a whole that is not derived through considerable of its parts"), in which regions are recognized and boundaries drawn intuitively; a map overlay system where different layers like geology, landforms and soil types are overlain to identify ecosystems; multivariate clustering of site attributes; digital image processing of remotely sensed data grouping areas based on their appearance or other spectral properties; or by a "controlling factors method" where a subset of factors (like soils, climate, vegetation physiognomy or the distribution of plant or animal species) are selected from a large array of possible ones are used to delineate ecosystems. In contrast with Bailey's methodology, Puerto Rico ecologist Ariel Lugo and coauthors identified ten characteristics of an effective classification system: that it be based on georeferenced, quantitative data; that it should minimize subjectivity and explicitly identify criteria and assumptions; that it should be structured around the factors that drive ecosystem processes; that it should reflect the hierarchical nature of ecosystems; that it should be flexible enough to conform to the various scales at which ecosystem management operates; that it should be tied to reliable measures of climate so that it can "anticipat[e] global climate change; that it be applicable worldwide; that it should be validated against independent data; that it take into account the sometimes complex relationship between climate, vegetation and ecosystem functioning; and that it should be able to adapt and improve as new data become available".

Types

- Aquatic ecosystem
    - o Marine ecosystem
        - ☐ Large marine ecosystem
    - o Freshwater ecosystem
        - ☐ Lake ecosystem

- ☐ River ecosystem
- ☐ Wetland

- Terrestrial ecosystem
  - o Forest
  - o Littoral zone
  - o Riparian zone
  - o Subsurface lithoautotrophic microbial ecosystem
  - o Urban ecosystem
  - o Desert

A freshwater ecosystem in Gran Canaria, an island of the Canary Islands.

Anthropogenic Threats

As human populations and per capita consumption grow, so do the resource demands imposed on ecosystems and the impacts of the human ecological footprint. Natural resources are not invulnerable and infinitely available. The environmental impacts of anthropogenic actions, which are processes or materials derived from human activities, are becoming more apparent—air and water quality are increasingly compromised, oceans are being overfished, pests and diseases are extending beyond their historical boundaries, and deforestation is exacerbating flooding downstream. It has been reported that approximately 40–50% of Earth's ice-free land surface has been heavily transformed or degraded by anthropogenic activities, 66% of marine fisheries are either overexploited or at their limit, atmospheric CO_2 has increased more than 30% since the advent of industrialization, and nearly 25% of Earth's bird species have gone extinct in the last two thousand years. Society is increasingly becoming aware that ecosystem services are not only limited, but also that they are threatened by human activities. The need to better consider long-term ecosystem health and its role in enabling human habitation and economic activity is urgent. To help inform decision-makers, many ecosystem services are being assigned economic values, often based on the cost of replacement with anthropogenic alternatives. The ongoing challenge of prescribing economic value to nature, for example through biodiversity banking, is prompting transdisciplinary shifts in how we recognize and manage the environment, social responsibility, business opportunities, and our future as a species.

Types of Ecosystem

Aquatic Ecosystem

An estuary mouth and coastal waters, part of an aquatic ecosystem

An aquatic ecosystem is an ecosystem in a body of water. Communities of organisms that are dependent on each other and on their environment live in aquatic ecosystems. The two main types of aquatic ecosystems are marine ecosystems and freshwater ecosystems.

Types

Marine

Marine ecosystems cover approximately 71% of the Earth's surface and contain approximately 97% of the planet's water. They generate 32% of the world's net primary production. They are distinguished from freshwater ecosystems by the presence of dissolved compounds, especially salts, in the water. Approximately 85% of the dissolved materials in seawater are sodium and chlorine. Seawater has an average salinity of 35 parts per thousand (ppt) of water. Actual salinity varies among different marine ecosystems.

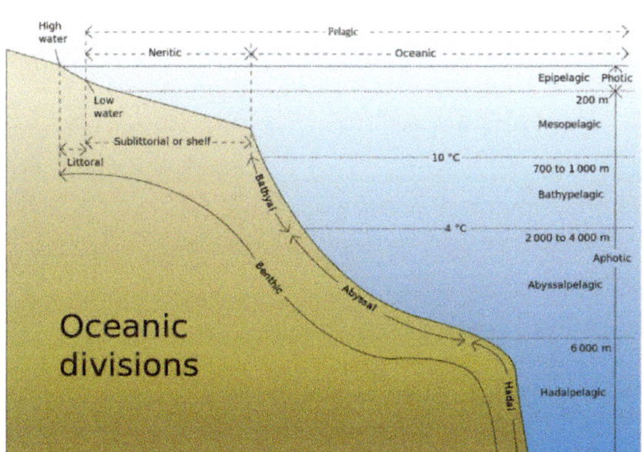

A classification of marine habitats.

Marine ecosystems can be divided into many zones depending upon water depth and shoreline features. The oceanic zone is the vast open part of the ocean where animals such as whales, sharks, and tuna live. The benthic zone consists of substrates below water where many invertebrates live. The intertidal zone is the area between high and low tides; in this figure it is termed the littoral zone. Other near-shore (neritic) zones can include estuaries, salt marshes, coral reefs, lagoons and mangrove swamps. In the deep water, hydrothermal vents may occur where chemosynthetic sulfur bacteria form the base of the food web.

Classes of organisms found in marine ecosystems include brown algae, dinoflagellates, corals, cephalopods, echinoderms, and sharks. Fishes caught in marine ecosystems are the biggest source of commercial foods obtained from wild populations.

Environmental problems concerning marine ecosystems include unsustainable exploitation of marine resources (for example overfishing of certain species), marine pollution, climate change, and building on coastal areas.

Freshwater

Freshwater ecosystems cover 0.78% of the Earth's surface and inhabit 0.009% of its total water. They generate nearly 3% of its net primary production. Freshwater ecosystems contain 41% of the world's known fish species.

There are three basic types of freshwater ecosystems:

- Lentic: slow moving water, including pools, ponds, and lakes.
- Lotic: faster moving water, for example streams and rivers.
- Wetlands: areas where the soil is saturated or inundated for at least part of the time.

Lentic

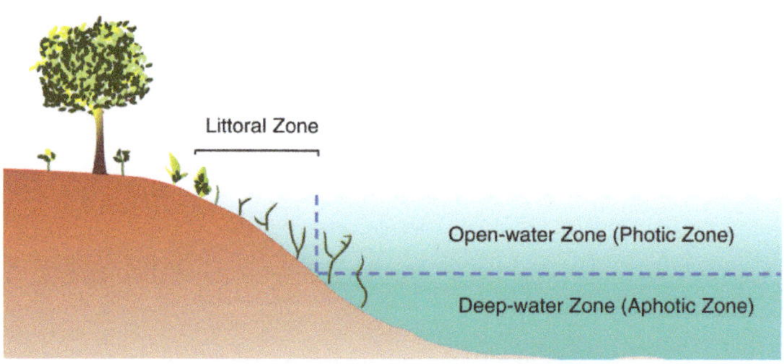

The three primary zones of a lake.

Lake ecosystems can be divided into zones. One common system divides lakes into three zones. The first, the littoral zone, is the shallow zone near the shore. This is where rooted wetland plants occur. The offshore is divided into two further zones, an open water zone and a deep water zone. In the open water zone (or photic zone) sunlight supports photosynthetic algae, and the species that feed upon them. In the deep water zone, sunlight is not available and the food web is based on detritus entering from the littoral and photic zones. Some systems use other names. The off shore areas may be called the pelagic zone, the photic zone may be called the limnetic zone and the aphotic zone may be called the profundal zone. Inland from the littoral zone one can also frequently identify a riparian zone which has plants still affected by the presence of the lake—this can include effects from windfalls, spring flooding, and winter ice damage. The production of the lake as a whole is the result of production from plants growing in the littoral zone, combined with production from plankton growing in the open water.

Wetlands can be part of the lentic system, as they form naturally along most lake shores, the width of the wetland and littoral zone being dependent upon the slope of the shoreline and the amount of natural change in water levels, within and among years. Often dead trees accumulate in this zone, either from windfalls on the shore or logs transported to the site during floods. This woody debris provides important habitat for fish and nesting birds, as well as protecting shorelines from erosion.

Two important subclasses of lakes are ponds, which typically are small lakes that intergrade with wetlands, and water reservoirs. Over long periods of time, lakes, or bays within them, may gradually become enriched by nutrients and slowly fill in with organic sediments, a process called succession. When humans use the watershed, the volumes of sediment entering the lake can accelerate this process. The addition of sediments and nutrients to a lake is known as eutrophication.

Ponds

Ponds are small bodies of freshwater with shallow and still water, marsh, and aquatic plants. They can be further divided into four zones: vegetation zone, open water, bottom mud and surface film. The size and depth of ponds often varies greatly with the time of year; many ponds are produced by spring flooding from rivers. Food webs are based both on free-floating algae and upon aquatic plants. There is usually a diverse array of aquatic life, with a few examples including algae, snails, fish, beetles, water bugs, frogs, turtles, otters and muskrats. Top predators may include large fish, herons, or alligators. Since fish are a major predator upon amphibian larvae, ponds that dry up each year, thereby killing resident fish, provide important refugia for amphibian breeding. Ponds that dry up completely each year are often known as vernal pools. Some ponds are produced by animal activity, including alligator holes and beaver ponds, and these add important diversity to landscapes.

Lotic

The major zones in river ecosystems are determined by the river bed's gradient or by the velocity of the current. Faster moving turbulent water typically contains greater concentrations of dissolved oxygen, which supports greater biodiversity than the slow moving water of pools. These distinctions form the basis for the division of rivers into upland and lowland rivers. The food base of streams within riparian forests is mostly derived from the trees, but wider streams and those that lack a canopy derive the majority of their food base from algae. Anadromous fish are also an important source of nutrients. Environmental threats to rivers include loss of water, dams, chemical pollution and introduced species. A dam produces negative effects that continue down the watershed. The most important negative effects are the reduction of spring flooding, which damages wetlands, and the retention of sediment, which leads to loss of deltaic wetlands.

Wetlands

Wetlands are dominated by vascular plants that have adapted to saturated soil. There are four main types of wetlands: swamp, marsh, fen and bog (both fens and bogs are types of mire). Wetlands are the most productive natural ecosystems in the world because of the proximity of water and soil. Hence they support large numbers of plant and animal species. Due to their productivity, wetlands are often converted into dry land with dykes and drains and used for agricultural purposes. The construction of dykes, and dams, has negative consequences for individual wetlands and entire watersheds. Their closeness to lakes and rivers means that they are often developed for human settlement. Once settlements are constructed and protected by dykes, the settlements then become vulnerable to land subsidence and ever increasing risk of flooding. The Louisiana coast around New Orleans is a well-known example; the Danube Delta in Europe is another.

Functions

Aquatic ecosystems perform many important environmental functions. For example, they recycle nutrients, purify water, attenuate floods, recharge ground water and provide habitats for wildlife. Aquatic ecosystems are also used for human recreation, and are very important to the tourism industry, especially in coastal regions.

The health of an aquatic ecosystem is degraded when the ecosystem's ability to absorb a stress has been exceeded. A stress on an aquatic ecosystem can be a result of physical, chemical or biological alterations of the environment. Physical alterations include changes in water temperature, water flow and light availability. Chemical alterations include changes in the loading rates of biostimulatory nutrients, oxygen consuming materials, and toxins. Biological alterations include over-harvesting of commercial species and the introduction of exotic species. Human populations can impose excessive stresses on aquatic ecosystems. There are many examples of excessive stresses with negative consequences. Consider three. The environmental history of the Great Lakes of North America illustrates this problem, particularly how multiple stresses, such as water pollution, over-harvesting and invasive species can combine. The Norfolk

Broadlands in England illustrate similar decline with pollution and invasive species. Lake Pontchartrain along the Gulf of Mexico illustrates the negative effects of different stresses including levee construction, logging of swamps, invasive species and salt water intrusion.

Abiotic Characteristics

An ecosystem is composed of biotic communities that are structured by biological interactions and abiotic environmental factors. Some of the important abiotic environmental factors of aquatic ecosystems include substrate type, water depth, nutrient levels, temperature, salinity, and flow. It is often difficult to determine the relative importance of these factors without rather large experiments. There may be complicated feedback loops. For example, sediment may determine the presence of aquatic plants, but aquatic plants may also trap sediment, and add to the sediment through peat.

The amount of dissolved oxygen in a water body is frequently the key substance in determining the extent and kinds of organic life in the water body. Fish need dissolved oxygen to survive, although their tolerance to low oxygen varies among species; in extreme cases of low oxygen some fish even resort to air gulping. Plants often have to produce aerenchyma, while the shape and size of leaves may also be altered. Conversely, oxygen is fatal to many kinds of anaerobic bacteria.

Nutrient levels are important in controlling the abundance of many species of algae. The relative abundance of nitrogen and phosphorus can in effect determine which species of algae come to dominate. Algae are a very important source of food for aquatic life, but at the same time, if they become over-abundant, they can cause declines in fish when they decay. Similar over-abundance of algae in coastal environments such as the Gulf of Mexico produces, upon decay, a hypoxic region of water known as a dead zone.

The salinity of the water body is also a determining factor in the kinds of species found in the water body. Organisms in marine ecosystems tolerate salinity, while many freshwater organisms are intolerant of salt. The degree of salinity in an estuary or delta is an important control upon the type of wetland (fresh, intermediate, or brackish), and the associated animal species. Dams built upstream may reduce spring flooding, and reduce sediment accretion, and may therefore lead to saltwater intrusion in coastal wetlands.

Freshwater used for irrigation purposes often absorbs levels of salt that are harmful to freshwater organisms.

Biotic Characteristics

The biotic characteristics are mainly determined by the organisms that occur. For example, wetland plants may produce dense canopies that cover large areas of sediment—or snails or geese may graze the vegetation leaving large mud flats. Aquatic environments have relatively low oxygen levels, forcing adaptation by the organisms found there. For example, many wetland plants must produce aerenchyma to carry oxygen to roots. Other biotic characteristics are more subtle and difficult to measure, such as the relative importance of competition, mutualism or predation. There are a growing number of cases where predation by coastal herbivores including snails, geese and mammals appears to be a dominant biotic factor.

Autotrophic Organisms

Autotrophic organisms are producers that generate organic compounds from inorganic material. Algae use solar energy to generate biomass from carbon dioxide and are possibly the most important autotrophic organisms in aquatic environments. Of course, the more shallow the water, the greater the biomass contribution from rooted and floating vascular plants. These two sources combine to produce the extraordinary production of estuaries and wetlands, as this autotrophic biomass is converted into fish, birds, amphibians and other aquatic species.

Chemosynthetic bacteria are found in benthic marine ecosystems. These organisms are able to feed on hydrogen sulfide in water that comes from volcanic vents. Great concentrations of animals that feed on these bacteria are found around volcanic vents. For example, there are giant tube worms (*Riftia pachyptila*) 1.5 m in length and clams (*Calyptogena magnifica*) 30 cm long.

Heterotrophic Organisms

Heterotrophic organisms consume autotrophic organisms and use the organic compounds in their bodies as energy sources and as raw materials to create their own biomass. Euryhaline organisms are salt tolerant and can survive in marine ecosystems, while stenohaline or salt intolerant species can only live in freshwater environments.

Marine Ecosystem

Coral reefs form complex marine ecosystems with tremendous biodiversity.

Marine ecosystems are among the largest of Earth's aquatic ecosystems, salt marshes, intertidal zones, estuaries, lagoons, mangroves, coral reefs, the deep sea, and the sea floor. They can be contrasted with freshwater ecosystems, which have a lower salt content. Marine waters cover two-thirds of the surface of the Earth. Such places are considered ecosystems because the plant life supports the animal life and vice versa.

Marine ecosystems are very important for the overall health of both marine and terrestrial environments. According to the World Resource Center, coastal habitats alone account for approximately 1/3 of all marine biological productivity, and estuarine ecosystems (i.e., salt marshes, seagrasses, mangrove forests) are among the most productive regions on the planet. In addition, other marine ecosystems such as coral reefs, provide food and shelter to the highest levels of marine diversity in the world.

Marine ecosystems usually have a large biodiversity and are therefore thought to have a good resistance against invasive species. However, exceptions have been observed, and the mechanisms responsible in determining the success of an invasion are not yet clear.

Freshwater Ecosystem

Freshwater angelfish

Freshwater ecosystems are a subset of Earth's aquatic ecosystems. They include lakes and ponds, rivers, streams, springs, and wetlands. They can be contrasted with marine ecosystems, which have a larger salt content. Freshwater habitats can be classified by different factors, including temperature, light penetration, and vegetation.

Freshwater ecosystems can be divided into lentic ecosystems (still water) and lotic ecosystems (flowing water).

Limnology (and its branch freshwater biology) is a study about freshwater ecosystems. It is a part of hydrobiology.

Original farts to understand and monitor freshwater ecosystems were spurred on by threats to human health (ex. Cholera outbreaks due to sewage contamination). Early monitoring focussed on chemical indicators, then bacteria, and finally algae, fungi and protozoa. A new type of monitoring involves differing groups of organisms (macroinvertebrates, macrophytes and fish) and the stream conditions associated with them.

Current biomonitoring techniques focus mainly on community structure or biochemical oxygen demand. Responses are measured by behavioural changes, altered rates of growth, reproduction or mortality. Macroinvertebrates are most often used in these models because of well known taxonomy, ease of collection, sensitivity to a range of stressors, and their overall value to the ecosystem. Most of these measurements are difficult to extrapolate on a large scale, however.

The use of reference sites is common when assessing what a healthy freshwater ecosystem should "look like". Reference sites are easier to reconstruct in standing water than moving water. Preserved indicators such as diatom valves, macrophyte pollen, insect chitin and fish scales can be used to establish a reference ecosystem representative of a time before large scale human disturbance.

Common chemical stresses on freshwater ecosystem health include acidification, eutrophication and copper and pesticide contamination.

Extinction of Freshwater Fauna

Over 123 freshwater fauna species have gone extinct in North America since 1900. Of North American freshwater species, an estimated 48.5% of mussels, 22.8% of gastropods, 32.7% of crayfishes, 25.9% of amphibians, and 21.3% of fishes are either endangered or threatened. Extinction rates of many species may increase severely into the next century because of invasive species, loss of keystone species, and species which are already functionally extinct. Projected extinction rates for freshwater animals are around five times greater than for land animals, and are comparable to the rates for rainforest communities. Recent extinction trends can be attributed largely to sedimentation, stream fragmentation, chemical and organic pollutants, dams, and invasive species.

Terrestrial Ecosystem

A terrestrial ecosystem is an ecosystem found only on landforms. Six primary terrestrial ecosystems exist: tundra, taiga, temperate deciduous forest, tropical rain forest, grassland and desert.

A community of organisms and their environment that occurs on the land masses of continents and islands. Terrestrial ecosystems are distinguished from aquatic ecosystems by the lower availability of water and the consequent importance of water as a limiting factor. Terrestrial ecosystems are characterized by greater temperature fluctuations on both a diurnal and seasonal basis than occur in aquatic ecosystems in similar climates. The availability of light is greater in terrestrial ecosystems than in aquatic ecosystems because the atmosphere is more transparent in land than in water. Gases are more available in terrestrial ecosystems than in aquatic ecosystems. Those gases include carbon dioxide that serves as a substrate for photosynthesis, oxygen that serves as a substrate in aerobic respiration, and nitrogen that serves as a substrate for nitrogen fixation. Terrestrial environments are segmented into a subterranean portion from which most water and ions are obtained, and an atmospheric portion from which gases are obtained and where the physical energy of light is transformed into the organic energy of carbon-carbon bonds through the process of photosynthesis.

Terrestrial ecosystems occupy 55,660,000 mi^2 (144,150,000 km^2), or 28.26% of Earth's surface. Although they are comparatively recent in the history of life (the first terrestrial organisms appeared in the Silurian period, about 425 million years ago) and occupy a much smaller portion of Earth's surface than marine ecosystems, terrestrial ecosystems have been a major site of adaptive radiation of both plants and animals. Major plant taxa in terrestrial ecosystems are members of the division *Magnoliophyta* (flowering plants), of which there are about 275,000 species, and the division *Pinophyta* (conifers), of which there are about 500 species. Members of the division *Bryophyta* (mosses and liverworts), of which there are about 24,000 species, are also important in some terrestrial ecosystems. Major animal taxa in terrestrial ecosystems include the classes *Insecta* (insects) with about 900,000 species, *Aves* (birds) with 8,500 species, and *Mammalia* (mammals) with approximately 4,100 species.

Organisms in terrestrial ecosystems have adaptations that allow them to obtain water when the entire body is no longer bathed in that fluid, means of transporting the water from limited sites of acquisition to the rest of the body, and means of preventing the evaporation of water from body surfaces. They also have traits that provide body support in the atmosphere, a much less buoyant medium than water, and other traits that render them capable of withstanding the extremes of temperature, wind, and humidity that characterize terrestrial ecosystems. Finally, the organisms in terrestrial ecosystems have evolved many methods of transporting gametes in environments where fluid flow is much less effective as a transport medium.

The organisms in terrestrial ecosystems are integrated into a functional unit by specific, dynamic relationships due to the coupled processes of energy and chemical flow. Those relationships can be summarized by schematic diagrams of trophic webs, which place organisms according to their feeding relationships. The base of the food web is occupied by green plants, which are the only organisms capable of utilizing the energy of the Sun and inorganic nutrients obtained from the soil to produce organic molecules. Terrestrial food webs can be broken into two segments based on the status of the plant material that enters them. *Grazing food webs* are associated with the consumption of living plant material by herbivores. Detritus food webs are associated with the consumption of dead plant material by detritivores. The relative importance of those two types of food webs varies considerably in different types of terrestrial ecosystems. Grazing food webs are more important in grasslands, where over half of net primary productivity may be consumed by herbivores. Detritus food webs are more important in forests, where less than 5% of net primary productivity may be consumed by herbivores.

There is one type of extensive terrestrial ecosystem due solely to human activities and eight types that are natural ecosystems. Those natural ecosystems reflect the variation of precipitation and temperature over Earth's surface. The smallest land areas are occupied by tundra and temperate grassland ecosystems, and the largest land area is occupied by tropical forest. The most productive ecosystems are temperate and tropical forests, and the least productive are deserts and tundras. Cultivated lands, which together with grasslands and savannas utilized for grazing are referred to as agroecosystems, are of intermediate extent and productivity. Because of both their areal extent and their high average productivity, tropical forests are the most productive of all terrestrial ecosystems, contributing 45% of total estimated net primary productivity on land.

Ecosystem Engineer

Beavers are the stereotypical ecosystem engineer because of the effects their dams have on channel flow, geomorphology, and ecology.

Kelp are autogenic ecosystem engineers, by building the necessary structure for kelp forests

An ecosystem engineer is any organism that creates, significantly modifies, maintains or destroys a habitat. These organisms can have a large impact on the species richness and landscape-level heterogeneity of an area. As a result, ecosystem engineers are important for maintaining the health and stability of the environment they are living in. Since all organisms impact the environment they live in in one way or another, it has been proposed that the term "ecosystem engineers" be used only for keystone species whose behavior very strongly affects other organisms

Types

Jones et al. identified two different types of ecosystem engineers:

Allogenic Engineers

Allogenic engineers modify the environment (biophysical) by mechanically changing living or nonliving materials from one form to another. Beavers are the original model for ecosystem engineers; in the process of clearcutting and damming, beavers alter their ecosystem extensively. The addition of a dam will change both the distribution and the abundance of many organisms in the area. Caterpillars are another example in that by creating shelters from leaves, they are also creating shelters for other organisms which may occupy them either simultaneously or subsequently. An additional example may be that of woodpeckers or other birds who create holes in trees for them to nest in. Once these birds are through with them, the holes are used by other species of birds or mammals for housing.

Autogenic Engineers

Autogenic engineers modify the environment by modifying themselves. Trees are a good example, because as they grow, their trunks and branches create habitats for other living things; these may include squirrels, birds or insects among others. In the tropics, lianas connect trees, which allow many animals to travel exclusively through the forest canopy.

Importance

Being able to identify ecosystem engineers in an environment can be important when looking at the influence these individuals may have over other organisms living in the same environment – especially in terms of resource availability.

The presence of some ecosystem engineers has been linked to higher species richness at the landscape level. By modifying the habitat, organisms like the beaver create more habitat heterogeneity and so can support species not found elsewhere. Thoughts may be that similar to other umbrella species by conserving an ecosystem engineer you may be able to protect the overall diversity of a landscape. Beavers have also been shown to maintain habitats in such a way as to protect the rare St. Francis' satyr butterfly and increase plant diversity.

Biodiversity may also be affected by ecosystem engineer's ability to increase the complexity of processes within an ecosystem, potentially allowing greater species richness and diversity in the local environments. As an example, beavers have the capacity to modify riparian forest and expand wetland habitats, which results in an increase of the diversity of the habitats by allowing a greater number of species to inhabit the landscape. Coral-reef habitats, created by the ecosystem engineer coral species, hold some of the highest abundances of aquatic species in the world.

Introduced Species as Ecosystem Engineers

Species are able to be transported across all parts of the world by humans or human-made vessels at boundless rates resulting in foreign ecosystem engineers changing the dynamics of species interactions and the possibility for engineering to occur in locations that would not have been accessible by engineers without the mediation by humans.

Introduced species, which may be invasive species, are often ecosystem engineers. Kudzu, a leguminous plant introduced to the southeast U.S., changes the distribution and number of animal and bird species in the areas it invades. It also crowds out native plant species. The zebra mussel is an ecosystem engineer in North America. By providing refuge from predators, it encourages the growth of freshwater invertebrates through increasing microhabitats. Light penetration into infected lakes also improves the ecosystem, resulting in an increase in algae. In contrast to the benefits some ecosystem engineers can cause, invasive species often have the reverse effect.

Gordon Dam

Humans as Ecosystem Engineers

Humans are thought to be one of the most dramatic ecosystem engineers. Niche construction has been prevalent since the earliest days of human activity. Through urban development, agricultural practices, logging, damming and mining, humans have changed the way they interact with the environment. This interaction is more studied in the field of human ecology.

Due to the complexity of many communities and ecosystems, restoration projects are often difficult. Ecosystem engineers have been proposed as a means to restore a given area to its previous state. While ideally these would all be natural agents, with today's level of development some form of human intervention may be necessary as well. In addition to being able to assist in restoration ecology, ecosystem engineers may be a helpful agent in invasive species management. New fields are developing which focus on restoring those ecosystems which have been disrupted or destroyed by human activities as well as developing ecosystems that are sustainable with both human and ecological values.

Beaver dam on Smilga

Examples

Terrestrial Environments

Besides the previously mentioned beaver acting as an ecosystem engineer, other terrestrial animals do the same. This may be through feeding habits, migration patterns or other behaviors that result in more permanent changes.

Research has suggested primates as ecosystem engineers as a result of their feeding strategies – frugivory and folivory – making them act as seed dispersers. As a whole primates are very abundant and feed on a large quantity of fruit that is then distributed around their territory. Elephants have also been designated ecosystem engineers as they cause very large changes to their environment whether it be through feeding, digging or migratory behavior.

Not only animals are ecosystem engineers. Fungi are able to connect regions that are distant from one another and translocate nutrients between them. Doing so they create nutritional niches for xylophagous invertebrates, supply trees with N translocated from previously predated animals or even form an "underground pipeline" that redistributes carbon between trees. Thus fungi are engineers controlling nutrient cycles in ecosystems.

Prairie dogs are another terrestrial form of allogenic ecosystem engineers due to the fact that the species has the ability to perform substantial modifications by burrowing and turning soil. They are able to influence soils and vegetation of the landscape while providing underground corridors for arthropods, avians, other small mammals, and reptiles. This has a positive effect on species richness and diversity of their habitats which results in the prairie dogs being labelled as keystone species.

Parrotfish

Marine Environments

In marine environments, filter feeders and plankton are ecosystem engineers because they alter turbidity and light penetration, controlling the depth at which photosynthesis can occur. This in turn limits the primary productivity of benthic and pelagic habitats and influences consumption patterns between trophic groups.

Another example of ecosystem engineers in marine environments would be Scleractinian corals as they create the framework for the habitat most coral-reef organisms depend on. Some ecosystem engineers such as coral have help maintaining their environment. Parrotfish often help maintain coral reefs as they feed on macroalgae that competes with the coral. As this relationship is mutually beneficial, a positive feedback cycle is formed between the two organisms, making them both responsible for creating and maintaining coral reef ecosystems.

Ecosystem Services

Humankind benefits in a multitude of ways from ecosystems. Collectively, these benefits are becoming known as ecosystem services. Ecosystem services are regularly involved in the provisioning of clean drinking water and the decomposition of wastes. While scientists and environmentalists have discussed ecosystem services implicitly for decades, the ecosystem services concept itself was popularized by the Millennium

Ecosystem Assessment (MA) in the early 2000s. This grouped ecosystem services into four broad categories: *provisioning*, such as the production of food and water; *regulating*, such as the control of climate and disease; *supporting*, such as nutrient cycles and crop pollination; and *cultural*, such as spiritual and recreational benefits. To help inform decision-makers, many ecosystem services are being assigned economic values.

Honey bee on Avocado crop. Pollination is just one type of ecosystem service

History

While the notion of human dependence on Earth's ecosystems reaches to the start of homo sapiens' existence, the term 'natural capital' was first coined by E.F. Schumacher in 1973 in his book *Small is Beautiful* . Recognition of how ecosystems could provide more complex services to mankind date back to at least Plato (c. 400 BC) who understood that deforestation could lead to soil erosion and the drying of springs. Modern ideas of ecosystem services probably began with Marsh in 1864 when he challenged the idea that Earth's natural resources are unbounded by pointing out changes in soil fertility in the Mediterranean. It was not until the late 1940s that three key authors – Henry Fairfield Osborn, Jr, William Vogt, and Aldo Leopold – promoted recognition of human dependence on the environment.

In 1956, Paul Sears drew attention to the critical role of the ecosystem in processing wastes and recycling nutrients. In 1970, Paul Ehrlich and Rosa Weigert called attention to "ecological systems" in their environmental science textbook and "the most subtle and dangerous threat to man's existence... the potential destruction, by man's own activities, of those ecological systems upon which the very existence of the human species depends".

The term "environmental services" was introduced in a 1970 report of the *Study of Critical Environmental Problems*, which listed services including insect pollination, fisheries, climate regulation and flood control. In following years, variations of the term were used, but eventually 'ecosystem services' became the standard in scientific literature.

The ecosystem services concept has continued to expand and includes socio-economic and conservation objectives, which are discussed below. A history of the concepts and terminology of ecosystem services as of 1997, can be found in Daily's book "Nature's Services: Societal Dependence on Natural Ecosystems".

Definition

Per the 2006 Millennium Ecosystem Assessment (MA), ecosystem services are "the benefits people obtain from ecosystems." The MA also delineated the four categories of ecosystem services—supporting, provisioning, regulating and cultural—discussed below.

By 2010, there had evolved various working definitions and descriptions of ecosystem services in the literature. To prevent double counting in ecosystem services audits, for instance, The Economics of Ecosystems and Biodiversity (TEEB) replaced "Supporting Services" in the MA with "Habitat Services" and "ecosystem functions," defined as "a subset of the interactions between ecosystem structure and processes that underpin the capacity of an ecosystem to provide goods and services."

Four Categories

Detritivores like this dung beetle help to turn animal wastes into organic material that can be reused by primary producers.

The Millennium Ecosystem Assessment (MA) report 2005 defines *Ecosystem services* as benefits people obtain from ecosystems and distinguishes four categories of ecosystem services, where the so-called supporting services are regarded as the basis for the services of the other three categories. The following lists represent the definition and samples of each according to the MA:

Supporting Services

Ecosystem services "that are necessary for the production of all other ecosystem services". These include services such as nutrient recycling, primary production and soil formation. These services make it possible for the ecosystems to provide services such as food supply, flood regulation and water purification.

Provisioning Services

"Products obtained from ecosystems"

- food (including seafood and game), crops, wild foods, and spices
- raw materials (including lumber, skins, fuel wood, organic matter, fodder, and fertilizer)
- genetic resources (including crop improvement genes, and health care)
- water
- minerals (including diatomite)
- medicinal resources (including pharmaceuticals, chemical models, and test and assay organisms)
- energy (hydropower, biomass fuels)
- ornamental resources (including fashion, handicraft, jewelry, pets, worship, decoration and souvenirs like furs, feathers, ivory, orchids, butterflies, aquarium fish, shells, etc.)

Regulating Services

"Benefits obtained from the regulation of ecosystem processes"

- carbon sequestration and climate regulation
- waste decomposition and detoxification
- purification of water and air
- pest and disease control

Cultural Services

"Nonmaterial benefits people obtain from ecosystems through spiritual enrichment, cognitive development, reflection, recreation, and aesthetic experiences"

- cultural (including use of nature as motif in books, film, painting, folklore, national symbols, architect, advertising, etc.)
- spiritual and historical (including use of nature for religious or heritage value or natural)

- recreational experiences (including ecotourism, outdoor sports, and recreation)
- science and education (including use of natural systems for school excursions, and scientific discovery)
- Therapeutic (including Ecotherapy, social forestry and animal assisted therapy)

There is discussion as to how the concept of cultural ecosystem services can be operationalized. A good review of approaches in landscape aesthetics, cultural heritage, outdoor recreation, and spiritual significance to define and assess cultural values of our environment so that they fit into the ecosystem services approach is given by Daniel et al. who vote for models that explicitly link ecological structures and functions with cultural values and benefits. There also is a fundamental critique of the concept of cultural ecosystem services that builds on three arguments:

1. Pivotal cultural values attaching to the natural/cultivated environment rely on an area's unique character that cannot be addressed by methods that use universal scientific parameters to determine ecological structures and functions.
2. If a natural/cultivated environment has symbolic meanings and cultural values the object of these values are not ecosystems but shaped lifeworldly phenomena like mountains, lakes, forests, and, mainly, symbolic landscapes.
3. Those cultural values do result not from properties produced by ecosystems but are the product of a specific way of seeing within the given cultural framework of symbolic experience.

Examples

The following examples illustrate the relationships between humans and natural ecosystems through the services derived from them:

- In New York City, where the quality of drinking water had fallen below standards required by the U.S. Environmental Protection Agency (EPA), authorities opted to restore the polluted Catskill Watershed that had previously provided the city with the ecosystem service of water purification. Once the input of sewage and pesticides to the watershed area was reduced, natural abiotic processes such as soil absorption and filtration of chemicals, together with biotic recycling via root systems and soil microorganisms, water quality improved to levels that met government standards. The cost of this investment in natural capital was estimated between $1–1.5 billion, which contrasted dramatically with the estimated $6–8 billion cost of constructing a water filtration plant plus the $300 million annual running costs.
- Pollination of crops by bees is required for 15-30% of U.S. food production; most large-scale farmers import non-native honey bees to provide this ser-

vice. One study reports that in California's agricultural region, it was found that wild bees alone could provide partial or complete pollination services or enhance the services provided by honey bees through behavioral interactions. However, intensified agricultural practices can quickly erode pollination services through the loss of species and those remaining are unable to compensate for the difference. The results of this study also indicate that the proportion of chaparral and oak-woodland habitat available for wild bees within 1–2 km of a farm can strongly stabilize and enhance the provision of pollination services, thereby providing a potential insurance policy for farmers of this region.

- In watersheds of the Yangtze River (China), spatial models for water flow through different forest habitats were created to determine potential contributions for hydroelectric power in the region. By quantifying the relative value of ecological parameters (vegetation-soil-slope complexes), researchers were able to estimate the annual economic benefit of maintaining forests in the watershed for power services to be 2.2 times that if it were harvested once for timber.

- In the 1980s, mineral water company Vittel (now a brand of Nestlé Waters) faced a critical problem. Nitrates and pesticides were entering the company's springs in northeastern France. Local farmers had intensified agricultural practices and cleared native vegetation that previously had filtered water before it seeped into the aquifer used by Vittel. This contamination threatened the company's right to use the "natural mineral water" label under French law. In response to this business risk, Vittel developed an incentive package for farmers to improve their agricultural practices and consequently reduce water pollution that had affected Vittel's product. For example, Vittel provided subsidies and free technical assistance to farmers in exchange for farmers' agreement to enhance pasture management, reforest catchments, and reduce the use of agrochemicals. This is an example of a Payment for ecosystem services program.

- It was counted that to plant 15 000 ha new woodland in UK, if we consider only the value of timber, it would cost £ 79 000 000 which is more than the benefit of £ 65 000 000. If, however, we include all other benefits the trees in lowland could provide (like soil stabilization, wind deflection, recreation wind defense, food production, air purification, carbon storage, wildlife habitat, fuel production, cooling, flood prevention), the costs will increase due to displacing the profitable farmland (would be around £ 231 000 000) but will be overweight by benefits of £ 546 000 000.

- In Europe, various projects are implemented in order to define the values of concrete ecosystems and to implement this concept into decision making process. For example, "LIFE Viva grass" project aims to due this with grasslands in Baltics.

Ecology

Understanding of ecosystem services requires a strong foundation in ecology, which describes the underlying principles and interactions of organisms and the environment. Since the scales at which these entities interact can vary from microbes to landscapes, milliseconds to millions of years, one of the greatest remaining challenges is the descriptive characterization of energy and material flow between them. For example, the area of a forest floor, the detritus upon it, the microorganisms in the soil and characteristics of the soil itself will all contribute to the abilities of that forest for providing ecosystem services like carbon sequestration, water purification, and erosion prevention to other areas within the watershed. Note that it is often possible for multiple services to be bundled together and when benefits of targeted objectives are secured, there may also be ancillary benefits – the same forest may provide habitat for other organisms as well as human recreation, which are also ecosystem services.

The complexity of Earth's ecosystems poses a challenge for scientists as they try to understand how relationships are interwoven among organisms, processes and their surroundings. As it relates to human ecology, a suggested research agenda for the study of ecosystem services includes the following steps:

1. identification of *ecosystem service providers* (*ESP*s) – species or populations that provide specific ecosystem services – and characterization of their functional roles and relationships;

2. determination of community structure aspects that influence how ESPs function in their natural landscape, such as compensatory responses that stabilize function and non-random extinction sequences which can erode it;

3. assessment of key environmental (abiotic) factors influencing the provision of services;

4. measurement of the spatial and temporal scales ESPs and their services operate on.

Recently, a technique has been developed to improve and standardize the evaluation of ESP functionality by quantifying the relative importance of different species in terms of their efficiency and abundance. Such parameters provide indications of how species respond to changes in the environment (i.e. predators, resource availability, climate) and are useful for identifying species that are disproportionately important at providing ecosystem services. However, a critical drawback is that the technique does not account for the effects of interactions, which are often both complex and fundamental in maintaining an ecosystem and can involve species that are not readily detected as a priority. Even so, estimating the functional structure of an ecosystem and combining it with information about individual species traits can help us understand the resilience of an ecosystem amidst environmental change.

Many ecologists also believe that the provision of ecosystem services can be stabilized with biodiversity. Increasing biodiversity also benefits the variety of ecosystem services available to society. Understanding the relationship between biodiversity and an ecosystem's stability is essential to the management of natural resources and their services.

Redundancy Hypothesis

The concept of ecological redundancy is sometimes referred to as *functional compensation* and assumes that more than one species performs a given role within an ecosystem. More specifically, it is characterized by a particular species increasing its efficiency at providing a service when conditions are stressed in order to maintain aggregate stability in the ecosystem. However, such increased dependence on a compensating species places additional stress on the ecosystem and often enhances its susceptibility to subsequent disturbance. The redundancy hypothesis can be summarized as "species redundancy enhances ecosystem resilience".

Another idea uses the analogy of rivets in an airplane wing to compare the exponential effect the loss of each species will have on the function of an ecosystem; this is sometimes referred to as *rivet popping*. If only one species disappears, the loss of the ecosystem's efficiency as a whole is relatively small; however if several species are lost, the system essentially collapses as an airplane wing would, were it to lose too many rivets. The hypothesis assumes that species are relatively specialized in their roles and that their ability to compensate for one another is less than in the redundancy hypothesis. As a result, the loss of any species is critical to the performance of the ecosystem. The key difference is the rate at which the loss of species affects total ecosystem function.

Portfolio Effect

A third explanation, known as the *portfolio effect*, compares biodiversity to stock holdings, where diversification minimizes the volatility of the investment, or in this case, the risk in stability of ecosystem services. This is related to the idea of *response diversity* where a suite of species will exhibit differential responses to a given environmental perturbation and therefore when considered together, they create a stabilizing function that preserves the integrity of a service.

Several experiments have tested these hypotheses in both the field and the lab. In ECOTRON, a laboratory in the UK where many of the biotic and abiotic factors of nature can be simulated, studies have focused on the effects of earthworms and symbiotic bacteria on plant roots. These laboratory experiments seem to favor the rivet hypothesis. However, a study on grasslands at Cedar Creek Reserve in Minnesota seems to support the redundancy hypothesis, as have many other field studies.

Economics

Sustainable urban drainage pond near housing in Scotland. The filtering and cleaning of surface and waste water by natural vegetation is a form of ecosystem service.

There are questions regarding the environmental and economic values of ecosystem services. Some people may be unaware of the environment in general and humanity's interrelatedness with the natural environment, which may cause misconceptions. Although environmental awareness is rapidly improving in our contemporary world, ecosystem capital and its flow are still poorly understood, threats continue to impose, and we suffer from the so-called 'tragedy of the commons'. Many efforts to inform decision-makers of current versus future costs and benefits now involve organizing and translating scientific knowledge to economics, which articulate the consequences of our choices in comparable units of impact on human well-being. An especially challenging aspect of this process is that interpreting ecological information collected from one spatial-temporal scale does not necessarily mean it can be applied at another; understanding the dynamics of ecological processes relative to ecosystem services is essential in aiding economic decisions. Weighting factors such as a service's irreplaceability or bundled services can also allocate economic value such that goal attainment becomes more efficient.

The economic valuation of ecosystem services also involves social communication and information, areas that remain particularly challenging and are the focus of many researchers. In general, the idea is that although individuals make decisions for any variety of reasons, trends reveal the aggregative preferences of a society, from which the economic value of services can be inferred and assigned. The six major methods for valuing ecosystem services in monetary terms are:

- Avoided cost: Services allow society to avoid costs that would have been incurred in the absence of those services (e.g. waste treatment by wetland habitats avoids health costs)
- Replacement cost: Services could be replaced with man-made systems (e.g. res-

toration of the Catskill Watershed cost less than the construction of a water purification plant)

- Factor income: Services provide for the enhancement of incomes (e.g. improved water quality increases the commercial take of a fishery and improves the income of fishers)
- Travel cost: Service demand may require travel, whose costs can reflect the implied value of the service (e.g. value of ecotourism experience is at least what a visitor is willing to pay to get there)
- Hedonic pricing: Service demand may be reflected in the prices people will pay for associated goods (e.g. coastal housing prices exceed that of inland homes)
- Contingent valuation: Service demand may be elicited by posing hypothetical scenarios that involve some valuation of alternatives (e.g. visitors willing to pay for increased access to national parks)

A peer-reviewed study published in 1997 estimated the value of the world's ecosystem services and natural capital to be between US$16–54 trillion per year, with an average of US$33 trillion per year. However, Salles (2011) indicates 'The total value of biodiversity is infinite, so having debate about what is the total value of nature is actually pointless because we can't live without it'.

Management and Policy

Although monetary pricing continues with respect to the valuation of ecosystem services, the challenges in policy implementation and management are significant and multitudinous. The administration of common pool resources is a subject of extensive academic pursuit. From defining the problems to finding solutions that can be applied in practical and sustainable ways, there is much to overcome. Considering options must balance present and future human needs, and decision-makers must frequently work from valid but incomplete information. Existing legal policies are often considered insufficient since they typically pertain to human health-based standards that are mismatched with necessary means to protect ecosystem health and services. To improve the information available, one suggestion has involved the implementation of an *Ecosystem Services Framework* (ESF), which integrates the biophysical and socio-economic dimensions of protecting the environment and is designed to guide institutions through multidisciplinary information and jargon, helping to direct strategic choices.

Novel and expedient methods are needed to deal with managing Earth's ecosystem services. Local to regional collective management efforts might be considered appropriate for services like crop pollination or resources like water. Another approach that has become increasingly popular over the last decade is the marketing of ecosystem services protection. Payment and trading of services is an emerging worldwide small-scale solution where one can acquire credits for activities such as sponsoring the protection of carbon sequestration sources or the restoration of ecosystem service providers. In some cases, banks for handling such credits have been established and conservation companies have even gone public on stock exchanges, defining an evermore parallel link with economic endeavors and opportunities for tying into social perceptions. However, crucial for implementation are clearly defined land rights, which is often lacking in many developing countries. In particular, many forest-rich developing countries suffering deforestation experience conflict between different forest stakeholders. In addition, concerns for such global transactions include inconsistent compensation for services or resources sacrificed elsewhere and misconceived warrants for irresponsible use. Another approach has been focused on protecting ecosystem service 'hotspots'. Recognition that the conservation of many ecosystem services aligns with more traditional conservation goals (i.e. biodiversity) has led to the suggested merging of objectives for maximizing their mutual success. This may be particularly strategic when employing networks that permit the flow of services across landscapes, and might also facilitate securing the financial means to protect services through a diversification of investors.

For example, in recent years there has been interest in the valuation of ecosystem services provided by shellfish production and restoration. A keystone species, low in the food chain, bivalve shellfish such as oysters support a complex community of species by performing a number of functions essential to the diverse array of species that surround them. There is also increasing recognition that some shellfish species may impact or control many ecological processes; so much so that they are included on the list of "ecosystem engineers"—organisms that physically, biologically or chemically modify the environment around them in ways that influence the health of other organisms. Many of the ecological functions and processes performed or affected by shellfish contribute to human well-being by providing a stream of valuable ecosystem services over time by filtering out particulate materials and potentially mitigating water quality issues by controlling excess nutrients in the water.

Ecosystem-Based Adaptation (EbA)

Ecosystem-Based Adaptation or EbA is an emerging strategy for community development and environmental management that seeks to use an ecosystem services framework to help communities adapt to the effects of climate change. The Convention on Biological Diversity currently defines Ecosystem-Based Adaptation as "the use of biodiversity and ecosystem services to help people adapt to the adverse effects of climate change", which includes the use of "sustainable management, conservation and restoration of ecosystems, as part of an overall adaptation strategy that takes into account the multiple social, economic and cultural co-benefits for local communities".

In 2001, the Millennium Ecosystem Assessment announced that humanity's impact on the natural world was increasing to levels never before seen, and that the degradation of the planet's ecosystems would become a major barrier to achieving the Millennium Development Goals. In recognition of this fact, Ecosystem-Based Adaptation seeks to use the restoration of ecosystems as a stepping-stone to improving the quality of life in communities experiencing the impacts of climate change. Specifically, this involves the restoration of ecosystems that provide the community with essential services, such as the provisioning of food and water and protection from storm surges and flooding. EbA interventions typically combine elements of both climate change mitigation and adaptation to global warming to help address the community's current and future needs.

Collaborative planning between scientists, policy makers, and community members is an essential element of Ecosystem-Based Adaptation. By drawing on the expertise of outside experts and local residents alike, EbA seeks to develop unique solutions to unique problems, rather than simply replicating past projects.

Estuarine and Coastal Ecosystem Services

Ecosystem services are defined as the gains acquired by humankind from surroundings ecosystems. Four different types of ecosystem services have been distinguished by the scientific body: regulating services, provisioning services, cultural services and supporting services. An ecosystem does not necessarily offer all four types of services simultaneously; but given the intricate nature of any ecosystem, it is usually assumed that humans benefit from a combination of these services. The services offered by diverse types of ecosystems (forests, seas, coral reefs, mangroves, etc.) differ in nature and in consequence. In fact, some services directly affect the livelihood of neighboring human populations (such as fresh water, food or aesthetic value, etc.) while other services affect general environmental conditions by which humans are indirectly impacted (such as climate change, erosion regulation or natural hazard regulation, etc.).

Estuarine and coastal ecosystems are both marine ecosystems. An estuary is defined as the area in which a river meets the sea or the ocean. The waters surrounding this area are predominantly salty waters or brackish waters; and the incoming river water is dynamically motioned by the tide. An estuary strip may be covered by populations of reed (or similar plants) and/or sandbanks (or similar form or land).

A coastal ecosystem occurs in areas where the sea or ocean waters meet the land.

Regulating Services

Regulating services are the "benefits obtained from the regulation of ecosystem processes". In the case of coastal and estuarine ecosystems, these services include climate regulation, waste treatment and disease control and natural hazard regulation.

- Climate Regulation

Both the biotic and abiotic ensembles of marine ecosystems play a role in climate regulation. They act as sponges when it comes to gases in the atmosphere, retaining large levels of CO_2 and other Green House Gases (methane and nitrous oxide). Marine plants also use CO_2 for photosynthesis purposes and help in reducing the atmospheric CO_2. The oceans and seas absorb the heat from the atmosphere and redistribute it through the means of water currents, and atmospheric processes, such as evaporation and the reflection of light allow for the cooling and warming of the overlying atmosphere. The ocean temperatures are thus imperative to the regulation of the atmospheric temperatures in any part of the world: "without the ocean, the Earth would be unbearably hot during the daylight hours and frigidly cold, if not frozen, at night".

- Waste Treatment & Disease Regulation

Another service offered by marine ecosystem is the treatment of wastes, thus helping in the regulation of diseases. Wastes can be diluted and detoxified through transport across marine ecosystems; pollutants are removed from the environment and stored, buried or recycled in marine ecosystems: "Marine ecosystems break down organic waste through microbial communities that filter water, reduce/limit the effects of eutrophication, and break down toxic hydrocarbons into their basic components such as carbon dioxide, nitrogen, phosphorus, and water". The fact that waste is diluted with large volumes of water and moves with water currents leads to the regulation of diseases and the reduction of toxics in seafood.

- Buffer Zones

Coastal and estuarine ecosystems act as buffer zones against natural hazards and environmental disturbances, such as floods, cyclones, tidal surges and storms. The role they play is to "[absorb] a portion of the impact and thus [lessen] its effect on the land". Wetlands, for example, and the vegetation it supports – trees, root mats, etc. – retain large amounts of water (surface water, snowmelt, rain, groundwater) and then slowly releases them back, decreasing the likeliness of floods. Mangrove forests protect coastal shorelines from tidal erosion or erosion by currents; a process that was studied after the 1999 cyclone that hit India. Villages that were surrounded with mangrove forests encountered less damages than other villages that weren't protected by mangroves.

Provisioning Services

Provisioning services consist of all "the products obtained from ecosystems". Marine ecosystems provide people with: wild & cultured seafood, fresh water, fiber & fuel and biochemical & genetic resources.

- Marine Products

Humans consume a large number of products originating from the seas, whether as a nutritious product or for use in other sectors: "More than one billion people worldwide, or one-sixth of the global population, rely on fish as their main source of animal protein. In 2000, marine and coastal fisheries accounted for 12 per cent of world food production". Fish and other edible marine products - primarily fish, shellfish, roe and seaweeds – constitute for populations living along the coast the main elements of the local cultural diets, norms and traditions. A very pertinent example would be Sushi, the national food of Japan, which consists mostly of different types of fish and seaweed.

- Fresh Water

Water bodies that are not highly concentrated in salts are referred to as 'fresh water' bodies. Fresh water may run through lakes, rivers and streams, to name a few; but it is most prominently found in the frozen state or as soil moisture or buried deep underground. Fresh water is not only important for the survival of humans, but also for the survival of all the existing species of animals, plants.

- Raw Materials

Marine creatures provide us with the raw materials needed for the manufacturing of clothing, building materials (lime extracted from coral reefs), ornamental items and personal-use items (luffas, art and jewelry): "The skin of marine mammals for clothing, gas deposits for energy production, lime (extracted from coral reefs) for building construction, and the timber of mangroves and coastal forests for shelter are some of the more familiar uses of marine organisms. Raw marine materials are utilized for non-essential goods as well, such as shells and corals in ornamental items". Humans have also referred to processes within marine environments for the production of renewable energy: using the power of waves – or tidal power – as a source of energy for the powering of a turbine, for example. Oceans and seas are used as sites for offshore oil and gas installations, offshore wind farms.

- Biochemical and Genetic Resources

Biochemical resources are compounds extracted from marine organisms for use in medicines, pharmaceuticals, cosmetics and other biochemical products. Genetic resources are the genetic information found in marine organisms that would later on be used for animal and plant breeding and for technological advances in the biological field. These resources are either directly taken out from an organism – such as fish oil as a source of omega3 –, or used as a model for innovative man-made products: "such as the construction of fiber optics technology based on the properties of sponges. [...] Compared to terrestrial products, marine-sourced products tend to be more highly bioactive, likely due to the fact that marine organisms have to retain their potency despite being diluted in the surrounding sea-water".

Cultural Services

Cultural services relate to the non-material world, as they benefit the benefit recreational, aesthetic, cognitive and spiritual activities, which are not easily quantifiable in monetary terms.

- Inspirational

Marine environments have been used by many as an inspiration for their works of art, music, architecture, traditions... Water environments are spiritually important as a lot of people view them as a means for rejuvenation and change of perspective. Many also consider the water as being a part of their personality, especially if they have lived near it since they were kids: they associate it to fond memories and past experiences. Living near water bodies for a long time results in a certain set of water activities that become a ritual in the lives of people and of the culture in the region.

- Recreation and Tourism

Sea sports are very popular among coastal populations: surfing, snorkeling, whale watching, kayaking, recreational fishing...a lot of tourists also travel to resorts close to the sea or rivers or lakes to be able to experience these activities, and relax near the water.

Beach accommodated into a recreational area.

- Science and Education

A lot can be learned from marine processes, environments and organisms – that could be implemented into our daily actions and into the scientific domain. Although much is still yet to still be known about the ocean world: "by the extraordinary intricacy and complexity of the marine environment and how it is influenced by large spatial scales, time lags, and cumulative effects".

Supporting Services

Supporting services are the services that allow for the other ecosystem services to be present. They have indirect impacts on humans that last over a long period of time. Several services can be considered as being both supporting services and regulating/cultural/provisioning services.

- Nutrient Cycling

"Nutrient cycling refers to the storage, cycling, and maintenance of nutrients by organisms and their associated processes". The ocean is a vast storage pool for these nutrients, such as carbon, nitrogen and phosphorus. The nutrients are absorbed by the basic organisms of the marine food web and are thus transferred from one organism to the other and from one ecosystem to the other. Nutrients are recycled through the life cycle of organisms as they die and decompose, releasing the nutrients into the neighboring environment. "The service of nutrient cycling eventually impacts all other ecosystem services as all living things require a constant supply of nutrients to survive".

- Biologically Mediated Habitats

Biologically mediated habitats are defined as being the habitats that living marine structures offer to other organisms. These need not to be designed for the sole purpose of serving as a habitat, but happen to become living quarters whilst growing naturally. For example, coral reefs and mangrove forests are home to numerous species of fish, seaweed and shellfish... The importance of these habitats is that they allow for interactions between different species, aiding the provisioning of marine goods and services. They are also very important for the growth at the early life stages of marine species (breeding and bursary spaces), as they serve as a food source and as a shelter from predators.

Coral and other living organisms serve as habitats for many marine species.

- Primary Production

Primary production refers to the production of organic matter, i.e., chemically bound energy, through processes such as photosynthesis and chemosynthesis. The organic matter produced by primary producers forms the basis of all food webs. Further, it generates oxygen, a molecule necessary for life.

Businessworld

Ecosystem services degradation can pose a number of risks to corporate performance as well as provide business opportunities through ecosystem restoration and enhancement. Risks and opportunities include:

- *Operational*
    - Risks such as higher costs for freshwater due to scarcity or lower output for hydroelectric facilities due to siltation
    - Opportunities such as increasing water-use efficiency or building an on-site wetland to circumvent the need for new water treatment infrastructure
- *Regulatory and legal*
    - Risks such as new fines, government regulations, or lawsuits from local communities that lose ecosystem services due to corporate activities
    - Opportunities such as engaging governments to develop policies and incentives to protect or restore ecosystems that provide services a company needs
- *Reputational*
    - Risks such as retail companies being targeted by nongovernmental organization campaigns for purchasing wood or paper from sensitive forests

    Opportunities such as implementing and communicating sustainable purchasing, operating, or investment practices in order to differentiate corporate brands.
- *Market and product*
    - Risks such as customers switching to other suppliers that offer products with lower ecosystem impacts or governments implementing new sustainable procurement policies
    - Opportunities such as launching new products and services that reduce customer impacts on ecosystems or participating in emerging markets for carbon sequestration and watershed protection other products
- *Financing*
    - Risks such as banks implementing more rigorous lending requirements for corporate loans
    - Opportunities such as banks offering more favorable loan terms or investors taking positions in companies supplying products and services that improve resource use efficiency or restore degraded ecosystems

Many companies are not fully aware of the extent of their dependence and impact on ecosystems and the possible ramifications. Likewise, environmental management systems and environmental due diligence tools are more suited to handle "traditional" issues of pollution and natural resource consumption. Most focus on environmental impacts, not dependence. Several newly developed tools and methodologies can help the private sector value and assess ecosystem services. These include Our Ecosystem, the Corporate Ecosystem Services Review (ESR), Artificial Intelligence for Ecosystem Services (ARIES), the Natural Value Initiative (NVI) and InVEST (Integrated Valuation of Ecosystem Services & Tradeoffs)

Land Use Change Decisions

Ecosystem services decisions require making complex choices at the intersection of ecology, technology, society and the economy. The process of making ecosystem services decisions must consider the interaction of many types of information, honor all stakeholder viewpoints, including regulatory agencies, proposal proponents, decision makers, residents, NGOs, and measure the impacts on all four parts of the intersection. These decisions are usually spatial, always multi-objective, and based on uncertain data, models, and estimates. Often it is the combination of the best science combined with the stakeholder values, estimates and opinions that drive the process.

One analytical study modeled the stakeholders as agents to support water resource management decisions in the Middle Rio Grande basin of New Mexico. This study focused on modeling the stakeholder inputs across a spatial decision, but ignored uncertainty. Another study used Monte Carlo methods to exercise econometric models of landowner decisions in a study of the effects of land-use change. Here the stakeholder inputs were modeled as random effects to reflect the uncertainty. A third study used a Bayesian decision support system to both model the uncertainty in the scientific information Bayes Nets and to assist collecting and fusing the input from stakeholders. This study was about siting wave energy devices off the Oregon Coast, but presents a general method for managing uncertain spatial science and stakeholder information in a decision making environment. Remote sensing data and analyses can be used to assess the health and extent of land cover classes that provide ecosystem services, which aids in planning, management, monitoring of stakeholders' actions, and communication between stakeholders.

In Baltic countries scientists, nature conservationists and local authorities are implementing integrated planning approach for grassland ecosystems. They are developing Integrated Planning Tool that will be based on GIS (geographic information system) technology and put online that will help for planners to choose the best grassland management solution for concrete grassland. It will look holistically at the processes in the countryside and help to find best grassland management solutions by taking into account both natural and socioeconomic factors of the particular site.

References

- Schoener, Thomas W. (2009). "Ecological Niche". In Simon A. Levin. The Princeton Guide to Ecology. Princeton: Princeton University Press. pp. 2–13. ISBN 978-0-691-12839-9.
- Barange M, Field JG, Harris RP, Eileen E, Hofmann EE, Perry RI and Werner F (2010) Marine Ecosystems and Global Change Oxford University Press. ISBN 978-0-19-955802-5.
- Davenport J (2008) Challenges to Marine Ecosystems: Proceedings of the 41st European Marine Biology Symposium Volume 202 of Developments in hydrobiology. ISBN 978-1-4020-8807-0.
- Levner E, Linkov I and Proth J (2005) Strategic management of marine ecosystems Springer. Volume 50 of NATO Science Series IV. ISBN 978-1-4020-3158-8
- Mann KH and Lazier JRN (2006) Dynamics of marine ecosystems: biological-physical interactions in the oceans Wiley-Blackwell. ISBN 978-1-4051-1118-8.
- National Research Council (US) (1996) Freshwater ecosystems: revitalizing educational programs in limnology National Academy Press. ISBN 0-309-05443-5.
- Christensen V and Pauly D (eds.) (1993) Trophic models of aquatic ecosystems The WorldFish Center, issue 26 of ICLARM Technical Reports, volume 26 of ICLARM conference proceedings. ISBN 9789711022846.
- Boyd IL, Wanless S and Camphuysen CJ (2006) Top predators in marine ecosystems: their role in monitoring and management Volume 12 of Conservation biology series. Cambridge University Press. ISBN 978-0-521-84773-5

4

Understanding Ecology

The study of the connections between organisms and their respective environments is known as ecology. This field involves biology, geography and Earth science. Some of the aspects of ecology elucidated in this chapter are biogenesis, abiogenesis, human ecology, biome, ecological niche and biomass. The topics explained in this section serve as a source to understand the theory of ecology.

Ecology

Ecology is the scientific analysis and study of interactions among organisms and their environment. It is an interdisciplinary field that includes biology, geography, and Earth science. Ecology includes the study of interactions organisms have with each other, other organisms, and with abiotic components of their environment. Topics of interest to ecologists include the diversity, distribution, amount (biomass), and number (population) of particular organisms, as well as cooperation and competition between organisms, both within and among ecosystems. Ecosystems are composed of dynamically interacting parts including organisms, the communities they make up, and the non-living components of their environment. Ecosystem processes, such as primary production, pedogenesis, nutrient cycling, and various niche construction activities, regulate the flux of energy and matter through an environment. These processes are sustained by organisms with specific life history traits, and the variety of organisms is called biodiversity. Biodiversity, which refers to the varieties of species, genes, and ecosystems, enhances certain ecosystem services.

Ecology is not synonymous with environment, environmentalism, natural history, or environmental science. It is closely related to evolutionary biology, genetics, and ethology. An important focus for ecologists is to improve the understanding of how biodiversity affects ecological function. Ecologists seek to explain:

- Life processes, interactions, and adaptations
- The movement of materials and energy through living communities
- The successional development of ecosystems
- The abundance and distribution of organisms and biodiversity in the context of the environment.

Ecology is a human science as well. There are many practical applications of ecology in conservation biology, wetland management, natural resource management (agroecology, agriculture, forestry, agroforestry, fisheries), city planning (urban ecology), community health, economics, basic and applied science, and human social interaction (human ecology). For example, the *Circles of Sustainability* approach treats ecology as more than the environment 'out there'. It is not treated as separate from humans. Organisms (including humans) and resources compose ecosystems which, in turn, maintain biophysical feedback mechanisms that moderate processes acting on living (biotic) and non-living (abiotic) components of the planet. Ecosystems sustain life-supporting functions and produce natural capital like biomass production (food, fuel, fiber, and medicine), the regulation of climate, global biogeochemical cycles, water filtration, soil formation, erosion control, flood protection, and many other natural features of scientific, historical, economic, or intrinsic value.

The word "ecology" ("Ökologie") was coined in 1866 by the German scientist Ernst Haeckel (1834–1919). Ecological thought is derivative of established currents in philosophy, particularly from ethics and politics. Ancient Greek philosophers such as Hippocrates and Aristotle laid the foundations of ecology in their studies on natural history. Modern ecology became a much more rigorous science in the late 19th century. Evolutionary concepts relating to adaptation and natural selection became the cornerstones of modern ecological theory.

Integrative Levels, Scope, and Scale of Organization

The scope of ecology contains a wide array of interacting levels of organization spanning micro-level (e.g., cells) to a planetary scale (e.g., biosphere) phenomena. Ecosystems, for example, contain abiotic resources and interacting life forms (i.e., individual organisms that aggregate into populations which aggregate into distinct ecological communities). Ecosystems are dynamic, they do not always follow a linear successional path, but they are always changing, sometimes rapidly and sometimes so slowly that it can take thousands of years for ecological processes to bring about certain successional stages of a forest. An ecosystem's area can vary greatly, from tiny to vast. A single tree is of little consequence to the classification of a forest ecosystem, but critically relevant to organisms living in and on it. Several generations of an aphid population can exist over the lifespan of a single leaf. Each of those aphids, in turn, support diverse bacterial communities. The nature of connections in ecological communities cannot be explained by knowing the details of each species in isolation, because the emergent pattern is neither revealed nor predicted until the ecosystem is studied as an integrated whole. Some ecological principles, however, do exhibit collective properties where the sum of the components explain the properties of the whole, such as birth rates of a population being equal to the sum of individual births over a designated time frame.

Hierarchical Ecology

System behaviors must first be arrayed into different levels of organization. Behaviors corresponding to higher levels occur at slow rates. Conversely, lower organizational levels exhibit rapid rates. For example, individual tree leaves respond rapidly to momentary changes in light intensity, CO_2 concentration, and the like. The growth of the tree responds more slowly and integrates these short-term changes.

O'Neill et al. (1986)

The scale of ecological dynamics can operate like a closed system, such as aphids migrating on a single tree, while at the same time remain open with regard to broader scale influences, such as atmosphere or climate. Hence, ecologists classify ecosystems hierarchically by analyzing data collected from finer scale units, such as vegetation associations, climate, and soil types, and integrate this information to identify emergent patterns of uniform organization and processes that operate on local to regional, landscape, and chronological scales.

To structure the study of ecology into a conceptually manageable framework, the biological world is organized into a nested hierarchy, ranging in scale from genes, to cells, to tissues, to organs, to organisms, to species, to populations, to communities, to ecosystems, to biomes, and up to the level of the biosphere. This framework forms a panarchy and exhibits non-linear behaviors; this means that "effect and cause are disproportionate, so that small changes to critical variables, such as the number of nitrogen fixers, can lead to disproportionate, perhaps irreversible, changes in the system properties."

Biodiversity

Biodiversity refers to the variety of life and its processes. It includes the variety of living organisms, the genetic differences among them, the communities and ecosystems in which they occur, and the ecological and evolutionary processes that keep them functioning, yet ever changing and adapting.

Noss & Carpenter (1994)

Biodiversity (an abbreviation of "biological diversity") describes the diversity of life from genes to ecosystems and spans every level of biological organization. The term has several interpretations, and there are many ways to index, measure, characterize, and represent its complex organization. Biodiversity includes species diversity, ecosystem diversity, and genetic diversity and scientists are interested in the way that this diversity affects the complex ecological processes operating at and among these respective levels. Biodiversity plays an important role in ecosystem services which by definition maintain and improve human quality of life. Preventing species extinctions is one way to preserve biodiversity and that goal rests on techniques that preserve genetic diversity, habitat and the ability for species to migrate. Conservation priorities and management techniques require different approaches and considerations to address the full ecological scope of biodiversity. Natural capital that supports populations is critical for maintaining ecosystem services and species migration (e.g., riverine fish runs and avian insect control) has been implicated as one mechanism by which those service losses are experienced. An understanding of biodiversity has practical applications for species and ecosystem-level conservation planners as they make management recommendations to consulting firms, governments, and industry.

Habitat

The habitat of a species describes the environment over which a species is known to occur and the type of community that is formed as a result. More specifically, "habitats can be defined as regions in environmental space that are composed of multiple dimensions, each representing a biotic or abiotic environmental variable; that is, any component or characteristic of the environment related directly (e.g. forage biomass and quality) or indirectly (e.g. elevation) to the use of a location by the animal." For example, a habitat might be an aquatic or terrestrial environment that can be further categorized as a montane or alpine ecosystem. Habitat shifts provide important evidence of competition in nature where one population changes relative to the habitats that most other individuals of the species occupy. For example, one population of a species of tropical lizards (*Tropidurus hispidus*) has a flattened body relative to the main populations that live in open savanna. The population that lives in an isolated rock outcrop hides in crevasses where its flattened body offers a selective advantage. Habitat shifts also occur in the developmental life history of amphibians, and in insects that transition from aquatic to terrestrial habitats. Biotope and habitat are sometimes used interchangeably, but the former applies to a community's environment, whereas the latter applies to a species' environment.

Additionally, some species are ecosystem engineers, altering the environment within a localized region. For instance, beavers manage water levels by building dams which improves their habitat in a landscape.

Niche

Termite mounds with varied heights of chimneys regulate gas exchange, temperature and other environmental parameters that are needed to sustain the internal physiology of the entire colony.

Definitions of the niche date back to 1917, but G. Evelyn Hutchinson made conceptual advances in 1957 by introducing a widely adopted definition: "the set of biotic and abiotic conditions in which a species is able to persist and maintain stable population sizes." The ecological niche is a central concept in the ecology of organisms and is sub-divided into the *fundamental* and the *realized* niche. The fundamental niche is the set of environmental conditions under which a species is able to persist. The realized niche is the set of environmental plus ecological conditions under which a species persists. The Hutchinsonian niche is defined more technically as a "Euclidean hyperspace whose *dimensions* are defined as environmental variables and whose *size* is a function of the number of values that the environmental values may assume for which an organism has *positive fitness*."

Biogeographical patterns and range distributions are explained or predicted through knowledge of a species' traits and niche requirements. Species have functional traits that are uniquely adapted to the ecological niche. A trait is a measurable property, phenotype, or characteristic of an organism that may influence its survival. Genes play an important role in the interplay of development and environmental expression of traits. Resident species evolve traits that are fitted to the selection pressures of their local environment. This tends to afford them a competitive advantage and discourages similarly adapted species from having an overlapping geographic range. The competitive exclusion principle states that two species cannot coexist indefinitely by living off the same limiting resource; one will always out-compete the other. When similarly adapted species overlap geographically, closer inspection reveals subtle ecological differences in their habitat or dietary requirements. Some models and empirical studies, however, suggest that disturbances can stabilize the co-evolution and shared niche occupancy of similar species inhabiting species-rich communities. The habitat plus the niche is called the ecotope, which is defined as the full range of environmental and biological variables affecting an entire species.

Niche Construction

Organisms are subject to environmental pressures, but they also modify their habitats. The regulatory feedback between organisms and their environment can affect conditions from local (e.g., a beaver pond) to global scales, over time and even after death, such as decaying logs or silica skeleton deposits from marine organisms. The process and concept of ecosystem engineering is related to niche construction, but the former relates only to the physical modifications of the habitat whereas the latter also considers the evolutionary implications of physical changes to the environment and the feedback this causes on the process of natural selection. Ecosystem engineers are defined as: "organisms that directly or indirectly modulate the availability of resources to other species, by causing physical state changes in biotic or abiotic materials. In so doing they modify, maintain and create habitats."

The ecosystem engineering concept has stimulated a new appreciation for the influence that organisms have on the ecosystem and evolutionary process. The term "niche construction" is more often used in reference to the under-appreciated feedback mechanisms of natural selection imparting forces on the abiotic niche. An example of natural selection through ecosystem engineering occurs in the nests of social insects, including ants, bees, wasps, and termites. There is an emergent homeostasis or homeorhesis in the structure of the nest that regulates, maintains and defends the physiology of the entire colony. Termite mounds, for example, maintain a constant internal temperature through the design of air-conditioning chimneys. The structure of the nests themselves are subject to the forces of natural selection. Moreover, a nest can survive over successive generations, so that progeny inherit both genetic material and a legacy niche that was constructed before their time.

Biome

Biomes are larger units of organization that categorize regions of the Earth's ecosystems, mainly according to the structure and composition of vegetation. There are different methods to define the continental boundaries of biomes dominated by different functional types of vegetative communities that are limited in distribution by climate, precipitation, weather and other environmental variables. Biomes include tropical rainforest, temperate broadleaf and mixed forest, temperate deciduous forest, taiga, tundra, hot desert, and polar desert. Other researchers have recently categorized other biomes, such as the human and oceanic microbiomes. To a microbe, the human body is a habitat and a landscape. Microbiomes were discovered largely through advances in molecular genetics, which have revealed a hidden richness of microbial diversity on the planet. The oceanic microbiome plays a significant role in the ecological biogeochemistry of the planet's oceans.

Biosphere

The largest scale of ecological organization is the biosphere: the total sum of ecosystems on the planet. Ecological relationships regulate the flux of energy, nutrients, and climate all the way up to the planetary scale. For example, the dynamic history of the planetary atmosphere's CO_2 and O_2 composition has been affected by the biogenic flux of gases coming from respiration and photosynthesis, with levels fluctuating over time in relation to the ecology and evolution of plants and animals. Ecological theory has also been used to explain self-emergent regulatory phenomena at the planetary scale: for example, the Gaia hypothesis is an example of holism applied in ecological theory. The Gaia hypothesis states that there is an emergent feedback loop generated by the metabolism of living organisms that maintains the core temperature of the Earth and atmospheric conditions within a narrow self-regulating range of tolerance.

Individual Ecology

Understanding traits of individual organisms helps explain patterns and processes at other levels of organization including populations, communities, and ecosystems. Several areas of ecology of evolution that focus on such traits are life history theory, ecophysiology, metabolic theory of ecology, and Ethology. Examples of such traits include features of an organisms life cycle such as age to maturity, life span, or metabolic costs of reproduction. Other traits may be related to structure, such as the spines of a cactus or dorsal spines of a bluegill sunfish, or behaviors such as courtship displays or pair bonding. Other traits include emergent properties that are the result at least in part of interactions with the surrounding environment such as growth rate, resource uptake rate, winter, and deciduous vs. drought deciduous trees and shrubs.

One set of characteristics relate to body size and temperature. The metabolic theory of ecology provides a predictive qualitative set of relationships between an organism's body size and temperature and metabolic processes. In general, smaller, warmer organisms have higher metabolic rates and this results in a variety of predictions regarding individual somatic growth rates, reproduction and population growth rates, population size, and resource uptake rates.

The traits of organisms are subject to change through acclimation, development, and evolution. For this reason, individuals form a shared focus for ecology and for evolutionary ecology.

Population Ecology

Population ecology studies the dynamics of specie populations and how these populations interact with the wider environment. A population consists of individuals of the same species that live, interact, and migrate through the same niche and habitat.

A primary law of population ecology is the Malthusian growth model which states, "a population will grow (or decline) exponentially as long as the environment experienced by all individuals in the population remains constant." Simplified population models usually start with four variables: death, birth, immigration, and emigration.

An example of an introductory population model describes a closed population, such as on an island, where immigration and emigration does not take place. Hypotheses are evaluated with reference to a null hypothesis which states that random processes create the observed data. In these island models, the rate of population change is described by:

$$\frac{\mathrm{d}N}{\mathrm{d}T} = bN - dN = (b-d)N = rN,$$

where N is the total number of individuals in the population, b and d are the per capita rates of birth and death respectively, and r is the per capita rate of population change.

Using these modelling techniques, Malthus' population principle of growth was later transformed into a model known as the logistic equation:

$$\frac{dN}{dT} = aN\left(1 - \frac{N}{K}\right),$$

where N is the number of individuals measured as biomass density, a is the maximum per-capita rate of change, and K is the carrying capacity of the population. The formula states that the rate of change in population size (dN/dT) is equal to growth (aN) that is limited by carrying capacity ($1 - N/K$).

Population ecology builds upon these introductory models to further understand demographic processes in real study populations. Commonly used types of data include life history, fecundity, and survivorship, and these are analysed using mathematical techniques such as matrix algebra. The information is used for managing wildlife stocks and setting harvest quotas. In cases where basic models are insufficient, ecologists may adopt different kinds of statistical methods, such as the Akaike information criterion, or use models that can become mathematically complex as "several competing hypotheses are simultaneously confronted with the data."

Metapopulations and Migration

The concept of metapopulations was defined in 1969 as "a population of populations which go extinct locally and recolonize". Metapopulation ecology is another statistical approach that is often used in conservation research. Metapopulation models simplify the landscape into patches of varying levels of quality, and metapopulations are linked by the migratory behaviours of organisms. Animal migration is set apart from other kinds of movement; because, it involves the seasonal departure and return of individuals from a habitat. Migration is also a population-level phenomenon, as with the migration routes followed by plants as they occupied northern post-glacial environments. Plant ecologists use pollen records that accumulate and stratify in wetlands to reconstruct the timing of plant migration and dispersal relative to historic and contemporary climates. These migration routes involved an expansion of the range as plant populations expanded from one area to another. There is a larger taxonomy of movement, such as commuting, foraging, territorial behaviour, stasis, and ranging. Dispersal is usually distinguished from migration; because, it involves the one way permanent movement of individuals from their birth population into another population.

In metapopulation terminology, migrating individuals are classed as emigrants (when they leave a region) or immigrants (when they enter a region), and sites are classed either as sources or sinks. A site is a generic term that refers to places where ecologists sample populations, such as ponds or defined sampling areas in a forest. Source patches are productive sites that generate a seasonal supply of juveniles that migrate to other patch locations. Sink patches are unproductive sites that only receive migrants; the population at the site will disappear unless rescued by an adjacent source patch or environmental conditions become more favourable. Metapopulation models examine patch dynamics over time to answer potential questions about spatial and demographic ecology. The ecology of metapopulations is a dynamic process of extinction and colonization. Small patches of lower quality (i.e., sinks) are maintained or rescued by a seasonal influx of new immigrants. A dynamic metapopulation structure evolves from year to year, where some patches are sinks in dry years and are sources when conditions are more favourable. Ecologists use a mixture of computer models and field studies to explain metapopulation structure.

Community Ecology

Interspecific interactions such as predation are a key aspect of community ecology.

Community ecology examines how interactions among species and their environment affect the abundance, distribution and diversity of species within communities.

Johnson & Stinchcomb (2007)

Community ecology is the study of the interactions among a collections of species that inhabit the same geographic area. Community ecologists study the determinants of patterns and processes for two or more interacting species. Research in community ecology might measure species diversity in grasslands in relation to soil fertility. It might also include the analysis of predator-prey dynamics, competition among similar plant species, or mutualistic interactions between crabs and corals.

Ecosystem Ecology

These ecosystems, as we may call them, are of the most various kinds and sizes. They form one category of the multitudinous physical systems of the universe, which range from the universe as a whole down to the atom.

Tansley (1935)

Ecosystems may be habitats within biomes that form an integrated whole and a dynamically responsive system having both physical and biological complexes. Ecosystem ecology is the science of determining the fluxes of materials (e.g. carbon, phosphorus) between different pools (e.g., tree biomass, soil organic material). Ecosystem ecologist attempt to determine the underlying causes of these fluxes. Research in ecosystem ecology might measure primary production (g C/m^2) in a wetland in relation to decomposition and consumption rates (g C/m^2/y). This requires an understanding of the community connections between plants (i.e., primary producers) and the decomposers (e.g., fungi and bacteria),

A riparian forest in the White Mountains, New Hampshire (USA), an example of ecosystem ecology

The underlying concept of ecosystem can be traced back to 1864 in the published work of George Perkins Marsh ("Man and Nature"). Within an ecosystem, organisms are linked to the physical and biological components of their environment to which they are adapted. Ecosystems are complex adaptive systems where the interaction of life processes form self-organizing patterns across different scales of time and space. Ecosystems are broadly categorized as terrestrial, freshwater, atmospheric, or marine. Differences stem from the nature of the unique physical environments that shapes the biodiversity within each. A more recent addition to ecosystem ecology are technoecosystems, which are affected by or primarily the result of human activity.

Food Webs

A food web is the archetypal ecological network. Plants capture solar energy and use it to synthesize simple sugars during photosynthesis. As plants grow, they accumulate nutrients and are eaten by grazing herbivores, and the energy is transferred through a chain of organisms by consumption. The simplified linear feeding pathways that move from a basal trophic species to a top consumer is called the food chain. The larger interlocking pattern of food chains in an ecological community creates a complex food web. Food webs are a type of concept map or a heuristic device that is used to illustrate and study pathways of energy and material flows.

Food webs are often limited relative to the real world. Complete empirical measurements are generally restricted to a specific habitat, such as a cave or a pond, and principles gleaned from food web microcosm studies are extrapolated to larger systems. Feeding relations require extensive investigations into the gut contents of organisms, which can be difficult to decipher, or stable isotopes can be used to trace the flow of nutrient diets and energy through a food web. Despite these limitations, food webs remain a valuable tool in understanding community ecosystems.

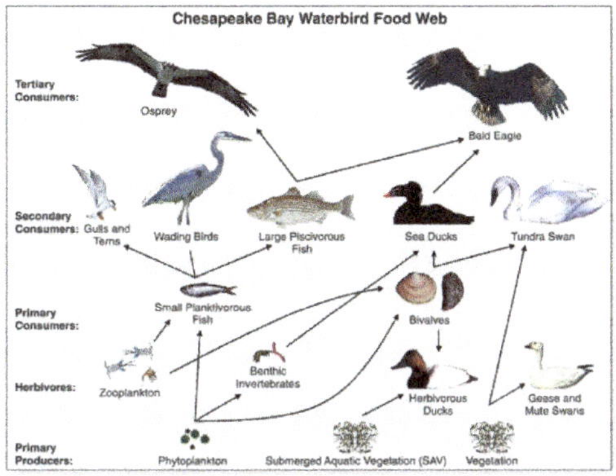

Generalized food web of waterbirds from Chesapeake Bay

Food webs exhibit principles of ecological emergence through the nature of trophic relationships: some species have many weak feeding links (e.g., omnivores) while some are more specialized with fewer stronger feeding links (e.g., primary predators). Theoretical and empirical studies identify non-random emergent patterns of few strong and many weak linkages that explain how ecological communities remain stable over time. Food webs are composed of subgroups where members in a community are linked by strong interactions, and the weak interactions occur between these subgroups. This increases food web stability. Step by step lines or relations are drawn until a web of life is illustrated.

Trophic Levels

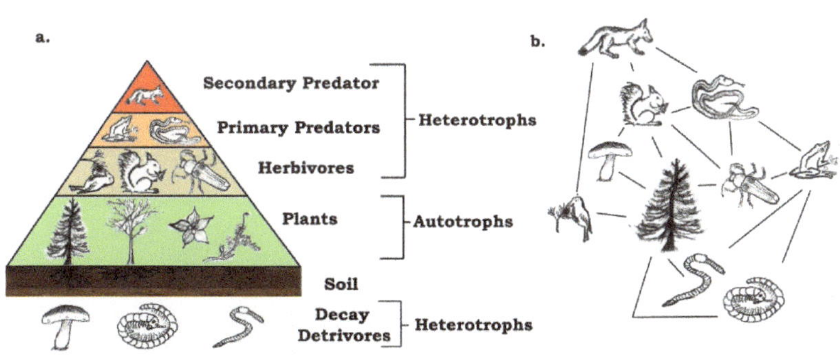

A trophic pyramid (a) and a food-web (b) illustrating ecological relationships among creatures that are typical of a northern boreal terrestrial ecosystem. The trophic pyramid roughly represents the biomass (usually measured as total dry-weight) at each level. Plants generally have the greatest biomass. Names of trophic categories are shown to the right of the pyramid. Some ecosystems, such as many wetlands, do not organize as a strict pyramid, because aquatic plants are not as productive as long-lived terrestrial plants such as trees. Ecological trophic pyramids are typically one of three kinds: 1) pyramid of numbers, 2) pyramid of biomass, or 3) pyramid of energy.

A trophic level is "a group of organisms acquiring a considerable majority of its energy from the adjacent level nearer the abiotic source." Links in food webs primarily connect feeding relations or trophism among species. Biodiversity within ecosystems can be organized into trophic pyramids, in which the vertical dimension represents feeding relations that become further removed from the base of the food chain up toward top predators, and the horizontal dimension represents the abundance or biomass at each level. When the relative abundance or biomass of each species is sorted into its respective trophic level, they naturally sort into a 'pyramid of numbers'.

Species are broadly categorized as autotrophs (or primary producers), heterotrophs (or consumers), and Detritivores (or decomposers). Autotrophs are organisms that produce their own food (production is greater than respiration) by photosynthesis or chemosynthesis. Heterotrophs are organisms that must feed on others for nourishment and energy (respiration exceeds production). Heterotrophs can be further sub-divided into different functional groups, including primary consumers (strict herbivores), secondary consumers (carnivorous predators that feed exclusively on herbivores), and tertiary consumers (predators that feed on a mix of herbivores and predators). Omnivores do not fit neatly into a functional category because they eat both plant and animal tissues. It has been suggested that omnivores have a greater functional influence as predators, because compared to herbivores, they are relatively inefficient at grazing.

Trophic levels are part of the holistic or complex systems view of ecosystems. Each trophic level contains unrelated species that are grouped together because they share common ecological functions, giving a macroscopic view of the system. While the notion of trophic levels provides insight into energy flow and top-down control within food webs, it is troubled by the prevalence of omnivory in real ecosystems. This has led some ecologists to "reiterate that the notion that species clearly aggregate into discrete, homogeneous trophic levels is fiction." Nonetheless, recent studies have shown that real trophic levels do exist, but "above the herbivore trophic level, food webs are better characterized as a tangled web of omnivores."

Keystone Species

Sea otters, an example of a keystone species

A keystone species is a species that is connected to a disproportionately large number of other species in the food-web. Keystone species have lower levels of biomass in the trophic pyramid relative to the importance of their role. The many connections that a keystone species holds means that it maintains the organization and structure of entire communities. The loss of a keystone species results in a range of dramatic cascading effects that alters trophic dynamics, other food web connections, and can cause the extinction of other species.

Sea otters (*Enhydra lutris*) are commonly cited as an example of a keystone species; because, they limit the density of sea urchins that feed on kelp. If sea otters are removed from the system, the urchins graze until the kelp beds disappear, and this has a dramatic effect on community structure. Hunting of sea otters, for example, is thought to have led indirectly to the extinction of the Steller's sea cow (*Hydrodamalis gigas*). While the keystone species concept has been used extensively as a conservation tool, it has been criticized for being poorly defined from an operational stance. It is difficult to experimentally determine what species may hold a keystone role in each ecosystem. Furthermore, food web theory suggests that keystone species may not be common, so it is unclear how generally the keystone species model can be applied.

Ecological Complexity

Complexity is understood as a large computational effort needed to piece together numerous interacting parts exceeding the iterative memory capacity of the human mind. Global patterns of biological diversity are complex. This biocomplexity stems from the interplay among ecological processes that operate and influence patterns at different scales that grade into each other, such as transitional areas or ecotones spanning landscapes. Complexity stems from the interplay among levels of biological organization as energy, and matter is integrated into larger units that superimpose onto the smaller parts. "What were wholes on one level become parts on a higher one." Small scale patterns do not necessarily explain large scale phenomena, otherwise captured in the expression (coined by Aristotle) 'the sum is greater than the parts'.

"Complexity in ecology is of at least six distinct types: spatial, temporal, structural, process, behavioral, and geometric." From these principles, ecologists have identified emergent and self-organizing phenomena that operate at different environmental scales of influence, ranging from molecular to planetary, and these require different explanations at each integrative level. Ecological complexity relates to the dynamic resilience of ecosystems that transition to multiple shifting steady-states directed by random fluctuations of history. Long-term ecological studies provide important track records to better understand the complexity and resilience of ecosystems over longer temporal and broader spatial scales. These studies are managed by the International Long Term Ecological Network (LTER). The longest experiment in existence is the Park Grass Experiment, which was initiated in 1856. Another example is the Hubbard Brook study, which has been in operation since 1960.

Holism

Holism remains a critical part of the theoretical foundation in contemporary ecological studies. Holism addresses the biological organization of life that self-organizes into layers of emergent whole systems that function according to non-reducible properties. This means that higher order patterns of a whole functional system, such as an ecosystem, cannot be predicted or understood by a simple summation of the parts. "New properties emerge because the components interact, not because the basic nature of the components is changed."

Ecological studies are necessarily holistic as opposed to reductionistic. Holism has three scientific meanings or uses that identify with ecology: 1) the mechanistic complexity of ecosystems, 2) the practical description of patterns in quantitative reductionist terms where correlations may be identified but nothing is understood about the causal relations without reference to the whole system, which leads to 3) a metaphysical hierarchy whereby the causal relations of larger systems are understood without reference to the smaller parts. Scientific holism differs from mysticism that has appropriated the same term. An example of metaphysical holism is identified in the trend of increased exterior thickness in shells of different species. The reason for a thickness increase can be understood through reference to principles of natural selection via predation without need to reference or understand the biomolecular properties of the exterior shells.

Relation to Evolution

Ecology and evolution are considered sister disciplines of the life sciences. Natural selection, life history, development, adaptation, populations, and inheritance are examples of concepts that thread equally into ecological and evolutionary theory. Morphological, behavioural, and genetic traits, for example, can be mapped onto evolutionary trees to study the historical development of a species in relation to their functions and roles in different ecological circumstances. In this framework, the analytical tools of ecologists and evolutionists overlap as they organize, classify, and investigate life through common systematic principals, such as phylogenetics or the Linnaean system of taxonomy. The two disciplines often appear together, such as in the title of the journal *Trends in Ecology and Evolution*. There is no sharp boundary separating ecology from evolution, and they differ more in their areas of applied focus. Both disciplines discover and explain emergent and unique properties and processes operating across different spatial or temporal scales of organization. While the boundary between ecology and evolution is not always clear, ecologists study the abiotic and biotic factors that influence evolutionary processes, and evolution can be rapid, occurring on ecological timescales as short as one generation.

Behavioural Ecology

All organisms can exhibit behaviours. Even plants express complex behaviour, including memory and communication. Behavioural ecology is the study of an organism's behaviour in its environment and its ecological and evolutionary implications. Ethology is the study of observable movement or behaviour in animals. This could include investigations of motile sperm of plants, mobile phytoplankton, zooplankton swimming toward the female egg, the cultivation of fungi by weevils, the mating dance of a salamander, or social gatherings of amoeba.

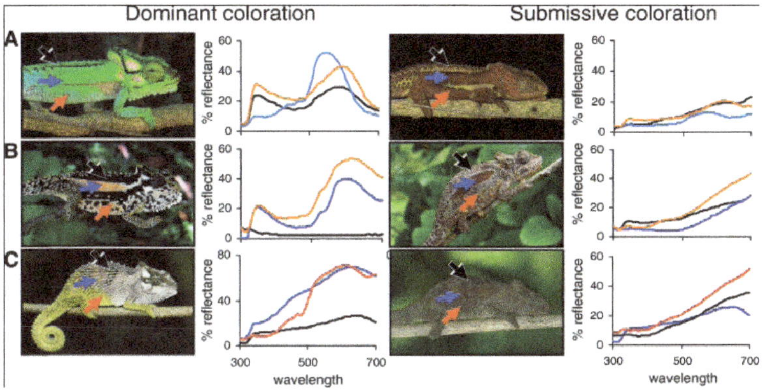

Social display and colour variation in differently adapted species of chameleons (*Bradypodion* spp.). Chameleons change their skin colour to match their background as a behavioural defence mechanism and also use colour to communicate with other members of their species, such as dominant (left) versus submissive (right) patterns shown in the three species (A-C) above.

Adaptation is the central unifying concept in behavioural ecology. Behaviours can be recorded as traits and inherited in much the same way that eye and hair colour can. Behaviours can evolve by means of natural selection as adaptive traits conferring functional utilities that increases reproductive fitness.

Predator-prey interactions are an introductory concept into food-web studies as well as behavioural ecology. Prey species can exhibit different kinds of behavioural adaptations to predators, such as avoid, flee, or defend. Many prey species are faced with multiple predators that differ in the degree of danger posed. To be adapted to their environment and face predatory threats, organisms must balance their energy budgets as they invest in different aspects of their life history, such as growth, feeding, mating, socializing, or modifying their habitat. Hypotheses posited in behavioural ecology are generally based on adaptive principles of conservation, optimization, or efficiency. For example, "[t]he threat-sensitive predator avoidance hypothesis predicts that prey should assess the degree of threat posed by different predators and match their behaviour according to current levels of risk" or "[t]he optimal flight initiation distance occurs where expected postencounter fitness is maximized, which depends on the prey's initial fitness, benefits obtainable by not fleeing, energetic escape costs, and expected fitness loss due to predation risk."

Elaborate sexual displays and posturing are encountered in the behavioural ecology of animals. The birds-of-paradise, for example, sing and display elaborate ornaments during courtship. These displays serve a dual purpose of signalling healthy or well-adapted individuals and desirable genes. The displays are driven by sexual selection as an advertisement of quality of traits among suitors.

Symbiosis: Leafhoppers (*Eurymela fenestrata*) are protected by ants (*Iridomyrmex purpureus*) in a symbiotic relationship. The ants protect the leafhoppers from predators and in return the leafhoppers feeding on plants exude honeydew from their anus that provides energy and nutrients to tending ants.

Cognitive Ecology

Cognitive ecology integrates theory and observations from evolutionary ecology and neurobiology, primarily cognitive science, in order to understand the effect that animal interaction with their habitat has on their cognitive systems and how those systems restrict behavior within an ecological and evolutionary framework. "Until recently, however, cognitive scientists have not paid sufficient attention to the fundamental fact that cognitive traits evolved under particular natural settings. With consideration of the selection pressure on cognition, cognitive ecology can contribute intellectual coherence to the multidisciplinary study of cognition." As a study involving the 'coupling' or interactions between organism and environment, cognitive ecology is closely related to enactivism, a field based upon the view that "...we must see the organism and environment as bound together in reciprocal specification and selection...".

Social Ecology

Social ecological behaviours are notable in the social insects, slime moulds, social spiders, human society, and naked mole-rats where eusocialism has evolved. Social behaviours include reciprocally beneficial behaviours among kin and nest mates and evolve from kin and group selection. Kin selection explains altruism through genetic relationships, whereby an altruistic behaviour leading to death is rewarded by the survival of genetic copies distributed among surviving relatives. The social insects, including ants, bees, and wasps are most famously studied for this type of relationship because the male drones are clones that share the same genetic make-up as every other male in the colony. In contrast, group selectionists find examples of altruism among non-genetic relatives and explain this through selection acting on the group; whereby, it becomes selectively advantageous for groups if their members express altruistic behaviours to one another. Groups with predominantly altruistic members beat groups with predominantly selfish members.

Coevolution

Bumblebees and the flowers they pollinate have coevolved so that both have become dependent on each other for survival.

Ecological interactions can be classified broadly into a host and an associate relationship. A host is any entity that harbours another that is called the associate. Relationships within a species that are mutually or reciprocally beneficial are called mutualisms. Examples of mutualism include fungus-growing ants employing agricultural symbiosis, bacteria living in the guts of insects and other organisms, the fig wasp and yucca moth pollination complex, lichens with fungi and photosynthetic algae, and corals with photosynthetic algae. If there is a physical connection between host and associate, the relationship is called symbiosis. Approximately 60% of all plants, for example, have a symbiotic relationship with arbuscular mycorrhizal fungi living in their roots forming an exchange network of carbohydrates for mineral nutrients.

Indirect mutualisms occur where the organisms live apart. For example, trees living in the equatorial regions of the planet supply oxygen into the atmosphere that sustains species living in distant polar regions of the planet. This relationship is called commensalism; because, many others receive the benefits of clean air at no cost or harm to trees supplying the oxygen. If the associate benefits while the host suffers, the relationship is called parasitism. Although parasites impose a cost to their host (e.g., via damage to their reproductive organs or propagules, denying the services of a beneficial partner), their net effect on host fitness is not necessarily negative and, thus, becomes difficult to forecast. Co-evolution is also driven by competition among species or among members of the same species under the banner of reciprocal antagonism, such as grasses competing for growth space. The Red Queen Hypothesis, for example, posits that parasites track down and specialize on the locally common genetic defense systems of its host that drives the evolution of sexual reproduction to diversify the genetic constituency of populations responding to the antagonistic pressure.

Parasitism: A harvestman arachnid being parasitized by mites. The harvestman is being consumed, while the mites benefit from traveling on and feeding off of their host.

Biogeography

Biogeography (an amalgamation of *biology* and *geography*) is the comparative study of the geographic distribution of organisms and the corresponding evolution of their traits in space and time. The *Journal of Biogeography* was established in 1974. Biogeography and ecology share many of their disciplinary roots. For example, the theory of island biogeography, published by the mathematician Robert MacArthur and ecologist Edward O. Wilson in 1967 is considered one of the fundamentals of ecological theory.

Biogeography has a long history in the natural sciences concerning the spatial distribution of plants and animals. Ecology and evolution provide the explanatory context for biogeographical studies. Biogeographical patterns result from ecological processes that influence range distributions, such as migration and dispersal. and from historical processes that split populations or species into different areas. The biogeographic processes that result in the natural splitting of species explains much of the modern distribution of the Earth's biota. The splitting of lineages in a species is called vicariance biogeography and it is a sub-discipline of biogeography. There are also practical applications in the field of biogeography concerning ecological systems and processes. For example, the range and distribution of biodiversity and invasive species responding to climate change is a serious concern and active area of research in the context of global warming.

r/K-Selection theory

A population ecology concept is r/K selection theory, one of the first predictive models in ecology used to explain life-history evolution. The premise behind the r/K selection model is that natural selection pressures change according to population density. For example, when an island is first colonized, density of individuals is low. The initial increase in population size is not limited by competition, leaving an abundance of available resources for rapid population growth. These early phases of population growth experience *density-independent* forces of natural selection, which is called *r*-selection. As the population becomes more crowded, it approaches the island's carrying capacity, thus forcing individuals to compete more heavily for fewer available resources. Under crowded conditions, the population experiences density-dependent forces of natural selection, called *K*-selection.

In the *r/K*-selection model, the first variable *r* is the intrinsic rate of natural increase in population size and the second variable *K* is the carrying capacity of a population. Different species evolve different life-history strategies spanning a continuum between these two selective forces. An *r*-selected species is one that has high birth rates, low levels of parental investment, and high rates of mortality before individuals reach maturity. Evolution favours high rates of fecundity in *r*-selected species. Many kinds of insects and invasive species exhibit *r*-selected characteristics. In contrast, a *K*-selected species has low rates of fecundity, high levels of parental investment in the young, and low rates of mortality as individuals mature. Humans and elephants are examples of species exhibiting *K*-selected characteristics, including longevity and efficiency in the conversion of more resources into fewer offspring.

Molecular Ecology

The important relationship between ecology and genetic inheritance predates modern techniques for molecular analysis. Molecular ecological research became more feasible with the development of rapid and accessible genetic technologies, such as the polymerase chain reaction (PCR). The rise of molecular technologies and influx of research questions into this new ecological field resulted in the publication *Molecular Ecology* in 1992. Molecular ecology uses various analytical techniques to study genes in an evolutionary and ecological context. In 1994, John Avise also played a leading role in this area of science with the publication of his book, *Molecular Markers, Natural History and Evolution*. Newer technologies opened a wave of genetic analysis into organisms once difficult to study from an ecological or evolutionary standpoint, such as bacteria, fungi, and nematodes. Molecular ecology engendered a new research paradigm for investigating ecological questions considered otherwise intractable. Molecular investigations revealed previously obscured details in the tiny intricacies of nature and improved resolution into probing questions about behavioural and biogeographical ecology. For example, molecular ecology revealed promiscuous sexual behaviour and multiple male partners in tree swallows previously thought to be socially monogamous. In a biogeographical context, the marriage between genetics, ecology, and evolution resulted in a new sub-discipline called phylogeography.

Human Ecology

The history of life on Earth has been a history of interaction between living things and their surroundings. To a large extent, the physical form and the habits of the earth's vegetation and its animal life have been molded by the environment. Considering the whole span of earthly time, the opposite effect, in which life actually modifies its surroundings, has been relatively slight. Only within the moment of time represented by the present century has one species man acquired significant power to alter the nature of his world.

Rachel Carson, "Silent Spring"

Ecology is as much a biological science as it is a human science. Human ecology is an interdisciplinary investigation into the ecology of our species. "Human ecology may be defined: (1) from a bio-ecological standpoint as the study of man as the ecological dominant in plant and animal communities and systems; (2) from a bio-ecological standpoint as simply another animal affecting and being affected by his physical environment; and (3) as a human being, somehow different from animal life in general, interacting with physical and modified environments in a distinctive and creative way. A truly interdisciplinary human ecology will most likely address itself to all three." The term was formally introduced in 1921, but many sociologists, geographers, psychologists, and other disciplines were interested in human relations to natural systems centuries prior, especially in the late 19th century.

The ecological complexities human beings are facing through the technological transformation of the planetary biome has brought on the Anthropocene. The unique set of circumstances has generated the need for a new unifying science called coupled human and natural systems that builds upon, but moves beyond the field of human ecology. Ecosystems tie into human societies through the critical and all encompassing life-supporting functions they sustain. In recognition of these functions and the incapability of traditional economic valuation methods to see the value in ecosystems, there has been a surge of interest in social-natural capital, which provides the means to put a value on the stock and use of information and materials stemming from ecosystem goods and services. Ecosystems produce, regulate, maintain, and supply services of critical necessity and beneficial to human health (cognitive and physiological), economies, and they even provide an information or reference function as a living library giving opportunities for science and cognitive development in children engaged in the complexity of the natural world. Ecosystems relate importantly to human ecology as they are the ultimate base foundation of global economics as every commodity, and the capacity for exchange ultimately stems from the ecosystems on Earth.

Restoration and Management

Ecosystem management is not just about science nor is it simply an extension of traditional resource management; it offers a fundamental reframing of how humans may work with nature.

Grumbine (1994)

Ecology is an employed science of restoration, repairing disturbed sites through human intervention, in natural resource management, and in environmental impact assessments. Edward O. Wilson predicted in 1992 that the 21st century "will be the era of restoration in ecology". Ecological science has boomed in the industrial investment of restoring ecosystems and their processes in abandoned sites after disturbance. Natural resource managers, in forestry, for example, employ ecologists to develop, adapt, and implement ecosystem based methods into the planning, operation, and restoration phases of land-use. Ecological science is used in the methods of sustainable harvesting, disease, and fire outbreak management, in fisheries stock management, for integrating land-use with protected areas and communities, and conservation in complex geo-political landscapes.

Relation to the Environment

The environment of ecosystems includes both physical parameters and biotic attributes. It is dynamically interlinked, and contains resources for organisms at any time throughout their life cycle. Like "ecology", the term "environment" has different conceptual meanings and overlaps with the concept of "nature". Environment "includes the physical world, the social world of human relations and the built world of human creation." The physical environment is external to the level of biological organization under investigation, including abiotic factors such as temperature, radiation, light, chemistry, climate and geology. The biotic environment includes genes, cells, organisms, members of the same species (conspecifics) and other species that share a habitat.

The distinction between external and internal environments, however, is an abstraction parsing life and environment into units or facts that are inseparable in reality. There is an interpenetration of cause and effect between the environment and life. The laws of thermodynamics, for example, apply to ecology by means of its physical state. With an understanding of metabolic and thermodynamic principles, a complete accounting of energy and material flow can be traced through an ecosystem. In this way, the environmental and ecological relations are studied through reference to conceptually manageable and isolated material parts. After the effective environmental components are understood through reference to their causes; however, they conceptually link back together as an integrated whole, or *holocoenotic* system as it was once called. This is known as the dialectical approach to ecology. The dialectical approach examines the parts, but integrates the organism and the environment into a dynamic whole (or umwelt). Change in one ecological or environmental factor can concurrently affect the dynamic state of an entire ecosystem.

Disturbance and Resilience

Ecosystems are regularly confronted with natural environmental variations and disturbances over time and geographic space. A disturbance is any process that removes biomass from a community, such as a fire, flood, drought, or predation. Disturbances occur over vastly different ranges in terms of magnitudes as well as distances and time periods, and are both the cause and product of natural fluctuations in death rates, species assemblages, and biomass densities within an ecological community. These disturbances create places of renewal where new directions emerge from the patchwork of natural experimentation and opportunity. Ecological resilience is a cornerstone theory in ecosystem management. Biodiversity fuels the resilience of ecosystems acting as a kind of regenerative insurance.

Metabolism and the Early Atmosphere

Metabolism – the rate at which energy and material resources are taken up from the environment, transformed within an organism, and allocated to maintenance, growth and reproduction – is a fundamental physiological trait.

Ernest et al.

The Earth was formed approximately 4.5 billion years ago. As it cooled and a crust and oceans formed, its atmosphere transformed from being dominated by hydrogen to one composed mostly of methane and ammonia. Over the next billion years, the metabolic activity of life transformed the atmosphere into a mixture of carbon dioxide, nitrogen, and water vapor. These gases changed the way that light from the sun hit the Earth's surface and greenhouse effects trapped heat. There were untapped sources of free energy within the mixture of reducing and oxidizing gasses that set the stage for primitive ecosystems to evolve and, in turn, the atmosphere also evolved.

The leaf is the primary site of photosynthesis in most plants.

Throughout history, the Earth's atmosphere and biogeochemical cycles have been in a dynamic equilibrium with planetary ecosystems. The history is characterized by periods of significant transformation followed by millions of years of stability. The evolution of the earliest organisms, likely anaerobic methanogen microbes, started the process by converting atmospheric hydrogen into methane ($4H_2 + CO_2 \rightarrow CH_4 + 2H_2O$). Anoxygenic photosynthesis reduced hydrogen concentrations and increased atmospheric methane, by converting hydrogen sulfide into water or other sulfur compounds (for example, $2H_2S + CO_2 + h\upsilon \rightarrow CH_2O + H_2O + 2S$). Early forms of fermentation also increased levels of atmospheric methane. The transition to an oxygen-dominant atmosphere (the *Great Oxidation*) did not begin until approximately 2.4–2.3 billion years ago, but photosynthetic processes started 0.3 to 1 billion years prior.

Radiation: Heat, Temperature and Light

The biology of life operates within a certain range of temperatures. Heat is a form of energy that regulates temperature. Heat affects growth rates, activity, behaviour, and primary production. Temperature is largely dependent on the incidence of solar radiation. The latitudinal and longitudinal spatial variation of temperature greatly affects climates and consequently the distribution of biodiversity and levels of primary production in different ecosystems or biomes across the planet. Heat and temperature relate importantly to metabolic activity. Poikilotherms, for example, have a body temperature that is largely regulated and dependent on the temperature of the external environment. In contrast, homeotherms regulate their internal body temperature by expending metabolic energy.

There is a relationship between light, primary production, and ecological energy budgets. Sunlight is the primary input of energy into the planet's ecosystems. Light is composed of electromagnetic energy of different wavelengths. Radiant energy from the sun generates heat, provides photons of light measured as active energy in the chemical reactions of life, and also acts as a catalyst for genetic mutation. Plants, algae, and some bacteria absorb light and assimilate the energy through photosynthesis. Organisms capable of assimilating energy by photosynthesis or through inorganic fixation of H_2S are autotrophs. Autotrophs — responsible for primary production — assimilate light energy which becomes metabolically stored as potential energy in the form of biochemical enthalpic bonds.

Physical Environments

Water

Wetland conditions such as shallow water, high plant productivity, and anaerobic substrates provide a suitable environment for important physical, biological, and chemical processes. Because of these processes, wetlands play a vital role in global nutrient and element cycles.

Cronk & Fennessy (2001)

Diffusion of carbon dioxide and oxygen is approximately 10,000 times slower in water than in air. When soils are flooded, they quickly lose oxygen, becoming hypoxic (an environment with O_2 concentration below 2 mg/liter) and eventually completely anoxic where anaerobic bacteria thrive among the roots. Water also influences the intensity and spectral composition of light as it reflects off the water surface and submerged particles. Aquatic plants exhibit a wide variety of morphological and physiological adaptations that allow them to survive, compete, and diversify in these environments. For example, their roots and stems contain large air spaces (aerenchyma) that regulate the efficient transportation of gases (for example, CO_2 and O_2) used in respiration and photosynthesis. Salt water plants (halophytes) have additional specialized adaptations, such as the development of special organs for shedding salt and osmoregulating their internal salt (NaCl) concentrations, to live in estuarine, brackish, or oceanic environments. Anaerobic soil microorganisms in aquatic environments use nitrate, manganese ions, ferric ions, sulfate, carbon dioxide, and some organic compounds; other microorganisms are facultative anaerobes and use oxygen during respiration when the soil becomes drier. The activity of soil microorganisms and the chemistry of the water reduces the oxidation-reduction potentials of the water. Carbon dioxide, for example, is reduced to methane (CH_4) by methanogenic bacteria. The physiology of fish is also specially adapted to compensate for environmental salt levels through osmoregulation. Their gills form electrochemical gradients that mediate salt excretion in salt water and uptake in fresh water.

Gravity

The shape and energy of the land is significantly affected by gravitational forces. On a large scale, the distribution of gravitational forces on the earth is uneven and influences the shape and movement of tectonic plates as well as influencing geomorphic processes such as orogeny and erosion. These forces govern many of the geophysical properties and distributions of ecological biomes across the Earth. On the organismal scale, gravitational forces provide directional cues for plant and fungal growth (gravitropism), orientation cues for animal migrations, and influence the biomechanics and size of animals. Ecological traits, such as allocation of biomass in trees during growth are subject to mechanical failure as gravitational forces influence the position and structure of branches and leaves. The cardiovascular systems of animals are functionally adapted to overcome pressure and gravitational forces that change according to the features of organisms (e.g., height, size, shape), their behaviour (e.g., diving, running, flying), and the habitat occupied (e.g., water, hot deserts, cold tundra).

Pressure

Climatic and osmotic pressure places physiological constraints on organisms, especially those that fly and respire at high altitudes, or dive to deep ocean depths. These constraints influence vertical limits of ecosystems in the biosphere, as organisms are physiologically sensitive and adapted to atmospheric and osmotic water pressure differences. For example, oxygen levels decrease with decreasing pressure and are a limiting factor for life at higher altitudes. Water transportation by plants is another important ecophysiological process affected by osmotic pressure gradients. Water pressure in the depths of oceans requires that organisms adapt to these conditions. For example, diving animals such as whales, dolphins, and seals are specially adapted to deal with changes in sound due to water pressure differences. Differences between hagfish species provide another example of adaptation to deep-sea pressure through specialized protein adaptations.

Wind and Turbulence

The architecture of the inflorescence in grasses is subject to the physical pressures of wind and shaped by the forces of natural selection facilitating wind-pollination (anemophily).

Turbulent forces in air and water affect the environment and ecosystem distribution, form and dynamics. On a planetary scale, ecosystems are affected by circulation patterns in the global trade winds. Wind power and the turbulent forces it creates can influence heat, nutrient, and biochemical profiles of ecosystems. For example, wind running over the surface of a lake creates turbulence, mixing the water column and influencing the environmental profile to create thermally layered zones, affecting how fish, algae, and other parts of the aquatic ecosystem are structured. Wind speed and turbulence also influence evapotranspiration rates and energy budgets in plants and animals. Wind speed, temperature and moisture content can vary as winds travel across different land features and elevations. For example, the westerlies come into contact with the coastal and interior mountains of western North America to produce a rain shadow on the leeward side of the mountain. The air expands and moisture condenses as the winds increase in elevation; this is called orographic lift and can cause precipitation. This environmental process produces spatial divisions in biodiversity, as species adapted to wetter conditions are range-restricted to the coastal mountain valleys and unable to migrate across the xeric ecosystems (e.g., of the Columbia Basin in western North America) to intermix with sister lineages that are segregated to the interior mountain systems.

Fire

Forest fires modify the land by leaving behind an environmental mosaic that diversifies the landscape into different seral stages and habitats of varied quality (left). Some species are adapted to forest fires, such as pine trees that open their cones only after fire exposure (right).

Plants convert carbon dioxide into biomass and emit oxygen into the atmosphere. By approximately 350 million years ago (the end of the Devonian period), photosynthesis had brought the concentration of atmospheric oxygen above 17%, which allowed combustion to occur. Fire releases CO_2 and converts fuel into ash and tar. Fire is a significant ecological parameter that raises many issues pertaining to its control and suppression. While the issue of fire in relation to ecology and plants has been recognized for a long time, Charles Cooper brought attention to the issue of forest fires in relation to the ecology of forest fire suppression and management in the 1960s.

Native North Americans were among the first to influence fire regimes by controlling their spread near their homes or by lighting fires to stimulate the production of herbaceous foods and basketry materials. Fire creates a heterogeneous ecosystem age and canopy structure, and the altered soil nutrient supply and cleared canopy structure opens new ecological niches for seedling establishment. Most ecosystems are adapted to natural fire cycles. Plants, for example, are equipped with a variety of adaptations to deal with forest fires. Some species (e.g., *Pinus halepensis*) cannot germinate until after their seeds have lived through a fire or been exposed to certain compounds from smoke. Environmentally triggered germination of seeds is called serotiny. Fire plays a major role in the persistence and resilience of ecosystems.

Soils

Soil is the living top layer of mineral and organic dirt that covers the surface of the planet. It is the chief organizing centre of most ecosystem functions, and it is of critical importance in agricultural science and ecology. The decomposition of dead organic matter (for example, leaves on the forest floor), results in soils containing minerals and nutrients that feed into plant production. The whole of the planet's soil ecosystems is called the pedosphere where a large biomass of the Earth's biodiversity organizes into trophic levels. Invertebrates that feed and shred larger leaves, for example, create smaller bits for smaller organisms in the feeding chain. Collectively, these organisms are the detritivores that regulate soil formation. Tree roots, fungi, bacteria, worms, ants, beetles, centipedes, spiders, mammals, birds, reptiles, amphibians, and other less familiar creatures all work to create the trophic web of life in soil ecosystems. Soils form composite phenotypes where inorganic matter is enveloped into the physiology of a whole community. As organisms feed and migrate through soils they physically displace materials, an ecological process called bioturbation. This aerates soils and stimulates heterotrophic growth and production. Soil microorganisms are influenced by and feed back into the trophic dynamics of the ecosystem. No single axis of causality can be discerned to segregate the biological from geomorphological systems in soils. Paleoecological studies of soils places the origin for bioturbation to a time before the Cambrian period. Other events, such as the evolution of trees and the colonization of land in the Devonian period played a significant role in the early development of ecological trophism in soils.

Biogeochemistry and Climate

Ecologists study and measure nutrient budgets to understand how these materials are regulated, flow, and recycled through the environment. This research has led to an understanding that there is global feedback between ecosystems and the physical parameters of this planet, including minerals, soil, pH, ions, water, and atmospheric gases. Six major elements (hydrogen, carbon, nitrogen, oxygen, sulfur, and phosphorus; H, C, N, O, S, and P) form the constitution of all biological macromolecules and feed into the Earth's geochemical processes. From the smallest scale of biology, the combined effect of billions upon billions of ecological processes amplify and ultimately regulate the biogeochemical cycles of the Earth. Understanding the relations and cycles mediated between these elements and their ecological pathways has significant bearing toward understanding global biogeochemistry.

The ecology of global carbon budgets gives one example of the linkage between biodiversity and biogeochemistry. It is estimated that the Earth's oceans hold 40,000 gigatonnes (Gt) of carbon, that vegetation and soil hold 2070 Gt, and that fossil fuel emissions are 6.3 Gt carbon per year. There have been major restructurings in these global carbon budgets during the Earth's history, regulated to a large extent by the ecology of the land. For example, through the early-mid Eocene volcanic outgassing, the oxidation of methane stored in wetlands, and seafloor gases increased atmospheric CO_2 (carbon dioxide) concentrations to levels as high as 3500 ppm.

In the Oligocene, from twenty-five to thirty-two million years ago, there was another significant restructuring of the global carbon cycle as grasses evolved a new mechanism of photosynthesis, C_4 photosynthesis, and expanded their ranges. This new pathway evolved in response to the drop in atmospheric CO_2 concentrations below 550 ppm. The relative abundance and distribution of biodiversity alters the dynamics between organisms and their environment such that ecosystems can be both cause and effect in relation to climate change. Human-driven modifications to the planet's ecosystems (e.g., disturbance, biodiversity loss, agriculture) contributes to rising atmospheric greenhouse gas levels. Transformation of the global carbon cycle in the next century is projected to raise planetary temperatures, lead to more extreme fluctuations in weather, alter species distributions, and increase extinction rates. The effect of global warming is already being registered in melting glaciers, melting mountain ice caps, and rising sea levels. Consequently, species distributions are changing along waterfronts and in continental areas where migration patterns and breeding grounds are tracking the prevailing shifts in climate. Large sections of permafrost are also melting to create a new mosaic of flooded areas having increased rates of soil decomposition activity that raises methane (CH_4) emissions. There is concern over increases in atmospheric methane in the context of the global carbon cycle, because methane is a greenhouse gas that is 23 times more effective at absorbing long-wave radiation than CO_2 on a 100-year time scale. Hence, there is a relationship between global warming, decomposition and respiration in soils and wetlands producing significant climate feedbacks and globally altered biogeochemical cycles.

History

Early Beginnings

Ecology has a complex origin, due in large part to its interdisciplinary nature. Ancient Greek philosophers such as Hippocrates and Aristotle were among the first to record observations on natural history. However, they viewed life in terms of essentialism, where species were conceptualized as static unchanging things while varieties were seen as aberrations of an idealized type. This contrasts against the modern understanding of ecological theory where varieties are viewed as the real phenomena of interest and having a role in the origins of adaptations by means of natural selection. Early conceptions of ecology, such as a balance and regulation in nature can be traced to Herodotus (died *c.* 425 BC), who described one of the earliest accounts of mutualism in his observation of "natural dentistry". Basking Nile crocodiles, he noted, would open their mouths to give sandpipers safe access to pluck leeches out, giving nutrition to the sandpiper and oral hygiene for the crocodile. Aristotle was an early influence on the philosophical development of ecology. He and his student Theophrastus made extensive observations on plant and animal migrations, biogeography, physiology, and on their behaviour, giving an early analogue to the modern concept of an ecological niche.

Ecological concepts such as food chains, population regulation, and productivity were first developed in the 1700s, through the published works of microscopist Antoni van Leeuwenhoek (1632–1723) and botanist Richard Bradley (1688?–1732). Biogeogra-

al development that are analogous to the developmental stages of an organism. The Clementsian paradigm was challenged by Henry Gleason, who stated that ecological communities develop from the unique and coincidental association of individual organisms. This perceptual shift placed the focus back onto the life histories of individual organisms and how this relates to the development of community associations.

The Clementsian superorganism theory was an overextended application of an idealistic form of holism. The term "holism" was coined in 1926 by Jan Christiaan Smuts, a South African general and polarizing historical figure who was inspired by Clements' superorganism concept. Around the same time, Charles Elton pioneered the concept of food chains in his classical book *Animal Ecology*. Elton defined ecological relations using concepts of food chains, food cycles, and food size, and described numerical relations among different functional groups and their relative abundance. Elton's 'food cycle' was replaced by 'food web' in a subsequent ecological text. Alfred J. Lotka brought in many theoretical concepts applying thermodynamic principles to ecology.

In 1942, Raymond Lindeman wrote a landmark paper on the trophic dynamics of ecology, which was published posthumously after initially being rejected for its theoretical emphasis. Trophic dynamics became the foundation for much of the work to follow on energy and material flow through ecosystems. Robert MacArthur advanced mathematical theory, predictions, and tests in ecology in the 1950s, which inspired a resurgent school of theoretical mathematical ecologists. Ecology also has developed through contributions from other nations, including Russia's Vladimir Vernadsky and his founding of the biosphere concept in the 1920s and Japan's Kinji Imanishi and his concepts of harmony in nature and habitat segregation in the 1950s. Scientific recognition of contributions to ecology from non-English-speaking cultures is hampered by language and translation barriers.

This whole chain of poisoning, then, seems to rest on a base of minute plants which must have been the original concentrators. But what of the opposite end of the food chain—the human being who, in probable ignorance of all this sequence of events, has rigged his fishing tackle, caught a string of fish from the waters of Clear Lake, and taken them home to fry for his supper?

Rachel Carson (1962)

Ecology surged in popular and scientific interest during the 1960–1970s environmental movement. There are strong historical and scientific ties between ecology, environmental management, and protection. The historical emphasis and poetic naturalistic writings advocating the protection of wild places by notable ecologists in the history of conservation biology, such as Aldo Leopold and Arthur Tansley, have been seen as far removed from urban centres where, it is claimed, the concentration of pollution and environmental degradation is located. Palamar (2008) notes an overshadowing by mainstream environmentalism of pioneering women in the early 1900s who fought for urban health ecology (then called euthenics) and brought about changes in environmental legislation. Women such as Ellen Swallow Richards and Julia Lathrop, among others, were precursors to the more popularized environmental movements after the 1950s.

In 1962, marine biologist and ecologist Rachel Carson's book *Silent Spring* helped to mobilize the environmental movement by alerting the public to toxic pesticides, such as DDT, bioaccumulating in the environment. Carson used ecological science to link the release of environmental toxins to human and ecosystem health. Since then, ecologists have worked to bridge their understanding of the degradation of the planet's ecosystems with environmental politics, law, restoration, and natural resources management.

Biogenesis

Biogenesis is the production of new living organisms or organelles. The law of biogenesis, attributed to Louis Pasteur, is the conclusion that complex living things come only from other living things, by reproduction (e.g. a spider lays eggs, which develop into spiders). That is, modern life does not arise from non-living material, which was the position held by spontaneous generation. A related statement is *Omnis cellula e cellula*, "all cells [are] from cells;" this conclusion is one of the central statements of cell theory.

Biogenesis and Abiogenesis

The term biogenesis was coined by Henry Charlton Bastian to mean the generation of a life form from nonliving materials, however, Thomas Henry Huxley chose the term abiogenesis and redefined biogenesis for life arising from preexisting life. The generation of life from non-living material is called abiogenesis, and occurred at least once in the history of the Earth, or in the history of the Universe, when life first arose.

The term *biogenesis* may also refer to biochemical processes of production in living organisms.

Spontaneous Generation and Its Disproof

The Ancient Greeks believed that living things could spontaneously come into being from nonliving matter, and that the goddess Gaia could make life arise spontaneously from stones – a process known as *Generatio spontanea*. Aristotle disagreed, but he still believed that creatures could arise from dissimilar organisms or from soil. Variations of this concept of spontaneous generation still existed as late as the 17th century, but towards the end of the 17th century, a series of observations and arguments began that eventually discredited such ideas. This advance in scientific understanding was met with much opposition, with personal beliefs and individual prejudices often obscuring the facts.

William Harvey (1578-1657) was an early proponent of all life beginning from an egg, *omne vivum ex ovo*. Francesco Redi, an Italian physician, proved as early as 1668 that higher forms of life did not originate spontaneously by demonstrating that maggots come from eggs of flies. but proponents of spontaneous generation claimed that this did not apply to microbes and continued to hold that these could arise spontaneously. Attempts to disprove the spontaneous generation of life from non-life continued in the early 19th century with observations and experiments by Franz Schulze and Theodor Schwann. In 1745, John Needham added chicken broth to a flask and boiled it. He then let it cool and waited. Microbes grew, and he proposed it as an example of spontaneous generation. In 1768, Lazzaro Spallanzani repeated Needham's experiment but removed all the air from the flask. No growth occurred. In 1854, Heinrich G. F. Schröder (1810–1885) and Theodor von Dusch, and in 1859, Schröder alone, repeated the Helmholtz filtration experiment and showed that living particles can be removed from air by filtering it through cotton-wool.

In 1864, Louis Pasteur finally announced the results of his scientific experiments. In a series of experiments similar to those performed earlier by Needham and Spallanzani, Pasteur demonstrated that life does not arise in areas that have not been contaminated by existing life. Pasteur's empirical results were summarized in the phrase *Omne vivum ex vivo*.

Abiogenesis

Precambrian stromatolites in the Siyeh Formation, Glacier National Park.
In 2002, a paper in the scientific journal *Nature* suggested that these 3.5 Ga (billion years) old geological formations contain fossilized cyanobacteria microbes.
This suggests they are evidence of one of the earliest known life forms on Earth.

Abiogenesis or biopoiesis or OoL (Origins of Life), is the natural process of life arising from non-living matter, such as simple organic compounds. It is thought to have occurred on Earth between 3.8 and 4.1 billion years ago. Abiogenesis is studied through a combination of laboratory experiments and extrapolation from the characteristics of modern organisms, and aims to determine how pre-life chemical reactions gave rise to life on Earth.

The study of abiogenesis involves geophysical, chemical, and biological considerations, with more recent approaches attempting a synthesis of all three. Many approaches investigate how self-replicating molecules, or their components, came into existence. It is generally thought that current life on Earth is descended from an RNA world, although RNA-based life may not have been the first life to have existed. The classic Miller–Urey experiment and similar research demonstrated that most amino acids, the basic chemical constituents of the proteins used in all living organisms, can be synthesized from inorganic compounds under conditions intended to replicate those of the early Earth. Various external sources of energy that may have triggered these reactions have been proposed, including lightning and radiation. Other approaches ("metabolism-first" hypotheses) focus on understanding how catalysis in chemical systems on the early Earth might have provided the precursor molecules necessary for self-replication. Complex organic molecules have been found in the Solar System and in interstellar space, and these molecules may have provided starting material for the development of life on Earth.

The panspermia hypothesis alternatively suggests that microscopic life was distributed to the early Earth by meteoroids, asteroids and other small Solar System bodies and that life may exist throughout the Universe. It is speculated that the biochemistry of life may have begun shortly after the Big Bang, 13.8 billion years ago, during a habitable epoch when the age of the universe was only 10 to 17 million years. The panspermia hypothesis therefore answers questions of where, not how, life came to be; it only postulates that life may have originated in a locale outside the Earth.

Nonetheless, Earth remains the only place in the Universe known to harbour life, and fossil evidence from the Earth supplies most studies of abiogenesis. The age of the Earth is about 4.54 billion years; the earliest undisputed evidence of life on Earth dates from at least 3.5 billion years ago, and possibly as early as the Eoarchean Era, after a geological crust started to solidify following the earlier molten Hadean Eon. Microbial mat fossils have been found in 3.48 billion-year-old sandstone in Western Australia. Other early physical evidence of biogenic substances includes graphite and possibly stromatolites discovered in 3.7 billion-year-old metasedimentary rocks in southwestern Greenland, as well as "remains of biotic life" found in 4.1 billion-year-old rocks in Western Australia. According to a scientist who commented on the study, "If life arose relatively quickly on Earth ... then it could be common in the universe."

Early Geophysical Conditions on Earth

The Hadean Earth is thought to have had a secondary atmosphere, formed through degassing of the rocks that accumulated from planetesimal impactors. At first, it was thought that the Earth's atmosphere consisted of hydrogen compounds—methane, ammonia and water vapour—and that life began under such reducing conditions, which are conducive to the formation of organic molecules. During its formation, the Earth lost a significant part of its initial mass, with a nucleus of the heavier rocky elements of the protoplanetary disk remaining. According to later models, suggested by study of ancient minerals, the atmosphere in the late Hadean period consisted largely of nitrogen and carbon dioxide, with smaller amounts of carbon monoxide, hydrogen, and sulfur compounds. As Earth lacked the gravity to hold any molecular hydrogen, this component of the atmosphere would have been rapidly lost during the Hadean period, along with the bulk of the original inert gases. The solution of carbon dioxide in water is thought to have made the seas slightly acidic, giving it a pH of about 5.5. The atmosphere at the time has been characterized as a "gigantic, productive outdoor chemical laboratory." It may have been similar to the mixture of gases released today by volcanoes, which still support some abiotic chemistry.

Oceans may have appeared first in the Hadean Eon, as soon as two hundred million years (200 Ma) after the Earth was formed, in a hot 100 °C (212 °F) reducing environment, and the pH of about 5.8 rose rapidly towards neutral. This has been supported by the dating of 4.404 Ga-old zircon crystals from metamorphosed quartzite of Mount Narryer in Western Australia, which are evidence that oceans and continental crust existed within 150 Ma of Earth's formation. Despite the likely increased volcanism and existence of many smaller tectonic "platelets," it has been suggested that between 4.4 and 4.3 Ga (billion year), the Earth was a water world, with little if any continental crust, an extremely turbulent atmosphere and a hydrosphere subject to intense ultraviolet (UV) light, from a T Tauri stage Sun, cosmic radiation and continued bolide impacts.

The Hadean environment would have been highly hazardous to modern life. Frequent collisions with large objects, up to 500 kilometres (310 mi) in diameter, would have been sufficient to sterilize the planet and vaporize the ocean within a few months of impact, with hot steam mixed with rock vapour becoming high altitude clouds that would completely cover the planet. After a few months, the height of these clouds would have begun to decrease but the cloud base would still have been elevated for about the next thousand years. After that, it would have begun to rain at low altitude. For another two thousand years, rains would slowly have drawn down the height of the clouds, returning the oceans to their original depth only 3,000 years after the impact event.

Earliest Biological Evidence for Life on Earth

The most commonly accepted location of the root of the tree of life is between a monophyletic domain Bacteria and a clade formed by Archaea and Eukaryota of what is referred to as the "traditional tree of life" based on several molecular studies starting with C. Woese. A very small minority of studies have concluded differently, namely that the root is in the Domain Bacteria, either in the phylum Firmicutes or that the phylum Chloroflexi is basal to a clade with Archaea+Eukaryotes and the rest of Bacteria as proposed by Thomas Cavalier-Smith.

The earliest life on Earth existed before 3.5 billion years ago, during the Eoarchean Era when sufficient crust had solidified following the molten Hadean Eon. Physical evidence has been found in biogenic graphite in 3.7 billion-year-old metasedimentary rocks from southwestern Greenland and microbial mat fossils found in 3.48 billion-year-old sandstone from Western Australia. Evidence of early life in rocks from Akilia Island, near the Isua supracrustal belt in southwestern Greenland, dating to 3.7 billion years ago have shown biogenic carbon isotopes. At Strelley Pool, in the Pilbarra region of Western Australia, compelling evidence of early life has been found in pyrite-bearing sandstone in a fossilized beach, that showed rounded tubular cells that oxidized sulfur by photosynthesis in the absence of oxygen. More recently, geochemists have found evidence that life likely existed on Earth at least 4.1 billion years ago — 300 million years earlier than previous research suggested.

In the earlier period between 3.8 and 4.1 Ga, changes in the orbits of the giant planets may have caused a heavy bombardment by asteroids and comets that pockmarked the Moon and the other inner planets (Mercury, Mars, and presumably Earth and Venus). This would likely have repeatedly sterilized the planet, had life appeared before that time. Geologically, the Hadean Earth would have been far more active than at any other time in its history. Studies of meteorites suggests that radioactive isotopes such as aluminium-26 with a half-life of 7.17×10^5 years, and potassium-40 with a half-life of 1.250×10^9 years, isotopes mainly produced in supernovae, were much more common. Coupled with internal heating as a result of gravitational sorting between the core and the mantle, there would have been a great deal of mantle convection, with the probable result of many smaller and much more active tectonic plates than now exist.

The time periods between such devastating environmental events give time windows for the possible origin of life in the early environments. A study by Kevin A. Maher and David J. Stevenson shows that if the deep marine hydrothermal setting provides a suitable site for the origin of life, then abiogenesis could have happened as early as 4.0 to 4.2 Ga, whereas if it occurred at the surface of the Earth, abiogenesis could only have occurred between 3.7 and 4.0 Ga.

In July 2016, scientists reported identifying a set of 355 genes from the Last Universal Common Ancestor (LUCA) of all organisms living on Earth. This research, published by William F. Martin, genetically sequenced 6.1 million protein coding genes from sequenced prokaryotic genomes of various phylogenic trees, identified 355 protein clusters from amongst 286,514 protein clusters, that were probably common to LUCA. The results "depict LUCA as anaerobic, CO_2-fixing, H_2-dependent with a Wood–Ljungdahl pathway, N_2-fixing and thermophilic. LUCA's biochemistry was replete with FeS clusters and radical reaction mechanisms. Its cofactors reveal dependence upon transition metals, flavins, S-adenosyl methionine, coenzyme A, ferredoxin, molybdopterin, corrins and selenium. Its genetic code required nucleoside modifications and S-adenosylmethionine-dependent methylations." The results depict methanogenic clostria as a basal clade in the 355 phylogenies examined, and suggest that LUCA inhabited an anaerobic hydrothermal vent setting in a geochemically active environment rich in H_2, CO_2 and iron.

Conceptual History

Spontaneous Generation

Belief in spontaneous generation of certain forms of life from non-living matter goes back to Aristotle and ancient Greek philosophy and continued to have support in Western scholarship until the 19th century. This belief was paired with a belief in heterogenesis, i.e., that one form of life derived from a different form (e.g., bees from flowers). Classical notions of spontaneous generation held that certain complex, living organisms are generated by decaying organic substances. According to Aristotle, it was a readily observable truth that aphids arise from the dew that falls on plants, flies from putrid matter, mice from dirty hay, crocodiles from rotting logs at the bottom of bodies of water, and so on. In the 17th century, people began to question such assumptions. In 1646, Sir Thomas Browne published his *Pseudodoxia Epidemica* (subtitled *Enquiries into Very many Received Tenets, and commonly Presumed Truths*), which was an attack on false beliefs and "vulgar errors." His contemporary, Alexander Ross, erroneously refuted him, stating: "To question this [Ed.: i.e., spontaneous generation], is to question Reason, Sense, and Experience: If he doubts of this, let him go to *Ægypt*, and there he will finde the fields swarming with mice begot of the mud of *Nylus*, to the great calamity of the Inhabitants."

In 1665, Robert Hooke published the first drawings of a microorganism. Hooke was followed in 1676 by Antonie van Leeuwenhoek, who drew and described microorganisms that are now thought to have been protozoa and bacteria. Many felt the existence of microorganisms was evidence in support of spontaneous generation, since microorganisms seemed too simplistic for sexual reproduction, and asexual reproduction through cell division had not yet been observed. Van Leeuwenhoek took issue with the ideas common at the time that fleas and lice could spontaneously result from putrefaction, and that frogs could likewise arise from slime. Using a broad range of experiments ranging from sealed and open meat incubation and the close study of insect reproduction he became, by the 1680s, convinced that spontaneous generation was incorrect.

The first experimental evidence against spontaneous generation came in 1668 when Francesco Redi showed that no maggots appeared in meat when flies were prevented from laying eggs. It was gradually shown that, at least in the case of all the higher and readily visible organisms, the previous sentiment regarding spontaneous generation was false. The alternative seemed to be biogenesis: that every living thing came from a pre-existing living thing.

In 1768, Lazzaro Spallanzani demonstrated that microbes were present in the air, and could be killed by boiling. In 1861, Louis Pasteur performed a series of experiments that demonstrated that organisms such as bacteria and fungi do not spontaneously appear in sterile, nutrient-rich media, but could only appear by invasion from without.

The belief that self-ordering by spontaneous generation was impossible begged for an alternative. By the middle of the 19th century, the theory of biogenesis had accumulated so much evidential support, due to the work of Pasteur and others, that the alternative theory of spontaneous generation had been effectively disproven. John Desmond Bernal, a pioneer in X-ray crystallography, suggested that earlier theories such as spontaneous generation were based upon an explanation that life was continuously created as a result of chance events.

Etymology

The term *biogenesis* is usually credited to either Henry Charlton Bastian or to Thomas Henry Huxley. Bastian used the term around 1869 in an unpublished exchange with John Tyndall to mean "life-origination or commencement". In 1870, Huxley, as new president of the British Association for the Advancement of Science, delivered an address entitled *Biogenesis and Abiogenesis*. In it he introduced the term *biogenesis* (with an opposite meaning to Bastian's) as well as *abiogenesis*:

> And thus the hypothesis that living matter always arises by the agency of pre-existing living matter, took definite shape; and had, henceforward, a right to be considered and a claim to be refuted, in each particular case, before the production of living matter in any other way could be admitted by careful reasoners. It will be necessary for me to refer to this hypothesis so frequently, that, to save circumlocution, I shall call it the hypothesis of *Biogenesis*; and I shall term the contrary doctrine–that living matter may be produced by not living matter–the hypothesis of *Abiogenesis*.

Subsequently, in the preface to Bastian's 1871 book, *The Modes of Origin of Lowest Organisms*, Bastian referred to the possible confusion with Huxley's usage and explicitly renounced his own meaning:

> A word of explanation seems necessary with regard to the introduction of the new term *Archebiosis*. I had originally, in unpublished writings, adopted the word *Biogenesis* to express the same meaning—viz., life-origination or commencement. But in the mean time the word *Biogenesis* has been made use of, quite independently, by a distinguished biologist [Huxley], who wished to make it bear a totally different meaning. He also introduced the word *Abiogenesis*.

I have been informed, however, on the best authority, that neither of these words can—with any regard to the language from which they are derived—be supposed to bear the meanings which have of late been publicly assigned to them. Wishing to avoid all needless confusion, I therefore renounced the use of the word *Biogenesis*, and being, for the reason just given, unable to adopt the other term, I was compelled to introduce a new word, in order to designate the process by which living matter is supposed to come into being, independently of pre-existing living matter.

Louis Pasteur and Charles Darwin

Charles Darwin in 1879

Louis Pasteur remarked, about a finding of his in 1864 which he considered definitive, "Never will the doctrine of spontaneous generation recover from the mortal blow struck by this simple experiment." One alternative was that life's origins on Earth had come from somewhere else in the Universe. Periodically resurrected Bernal said that this approach "is equivalent in the last resort to asserting the operation of metaphysical, spiritual entities... it turns on the argument of creation by design by a creator or demiurge." Such a theory, Bernal said, was unscientific. A theory popular around the same time was that life was the result of an inner "life force", which in the late 19th century was championed by Henri Bergson.

The idea of evolution by natural selection proposed by Charles Darwin put an end to these metaphysical theologies. In a letter to Joseph Dalton Hooker on 1 February 1871, Darwin discussed the suggestion that the original spark of life may have begun in a "warm little pond, with all sorts of ammonia and phosphoric salts, light, heat, electricity, &c., present, that a proteine compound was chemically formed ready to undergo still more complex changes." He went on to explain that "at the present day such matter would be instantly devoured or absorbed, which would not have been the case before living creatures were formed." He had written to Hooker in 1863 stating that, "It is mere rubbish, thinking at present of the origin of life; one might as well think of the origin of matter." In *On the Origin of Species*, he had referred to life having been "created", by which he "really meant 'appeared' by some wholly unknown process", but had soon regretted using the Old Testament term "creation".

"Primordial Soup" Hypothesis

Alexander Oparin (right) at the laboratory

No new notable research or theory on the subject appeared until 1924, when Alexander Oparin reasoned that atmospheric oxygen prevents the synthesis of certain organic compounds that are necessary building blocks for the evolution of life. In his book *The Origin of Life*, Oparin proposed that the "spontaneous generation of life" that had been attacked by Louis Pasteur did in fact occur once, but was now impossible because the conditions found on the early Earth had changed, and preexisting organisms would immediately consume any spontaneously generated organism. Oparin argued that a "primeval soup" of organic molecules could be created in an oxygenless atmosphere through the action of sunlight. These would combine in ever more complex ways until they formed coacervate droplets. These droplets would "grow" by fusion with other droplets, and "reproduce" through fission into daughter droplets, and so have a primitive metabolism in which factors that promote "cell integrity" survive, and those that do not become extinct. Many modern theories of the origin of life still take Oparin's ideas as a starting point.

Robert Shapiro has summarized the "primordial soup" theory of Oparin and J. B. S. Haldane in its "mature form" as follows:

1. The early Earth had a chemically reducing atmosphere.
2. This atmosphere, exposed to energy in various forms, produced simple organic compounds ("monomers").
3. These compounds accumulated in a "soup" that may have concentrated at various locations (shorelines, oceanic vents etc.).
4. By further transformation, more complex organic polymers - and ultimately life - developed in the soup.

About this time, Haldane suggested that the Earth's prebiotic oceans (quite different from their modern counterparts) would have formed a "hot dilute soup" in which organic compounds could have formed. Bernal called this idea *biopoiesis* or *biopoesis*, the process of living matter evolving from self-replicating but non-living molecules, and proposed that biopoiesis passes through a number of intermediate stages.

One of the most important pieces of experimental support for the "soup" theory came in 1952. Stanley L. Miller and Harold C. Urey performed an experiment that demonstrated how organic molecules could have spontaneously formed from inorganic precursors under conditions like those posited by the Oparin-Haldane hypothesis. The now-famous Miller–Urey experiment used a highly reducing mixture of gases - methane, ammonia, and hydrogen, as well as water vapour - to form basic organic monomers such as amino acids. The mixture of gases was cycled through an apparatus that delivered electrical sparks to the mixture. After one week, it was found that about 10% to 15% of the carbon in the system was now in the form of a racemic mixture of organic compounds, including amino acids, which are the building blocks of proteins. This provided direct experimental support for the second point of the "soup" theory, and it is around the remaining two points of the theory that much of the debate now centres.

Bernal shows that based upon this and subsequent work there is no difficulty in principle in forming most of the molecules we recognize as the basic molecules of life from their inorganic precursors. The underlying hypothesis held by Oparin, Haldane, Bernal, Miller and Urey, for instance, was that multiple conditions on the primeval Earth favoured chemical reactions that synthesized the same set of complex organic compounds from such simple precursors. A 2011 reanalysis of the saved vials containing the original extracts that resulted from the Miller and Urey experiments, using current and more advanced analytical equipment and technology, has uncovered more biochemicals than originally discovered in the 1950s. One of the more important findings was 23 amino acids, far more than the five originally found. However, Bernal said that "it is not enough to explain the formation of such molecules, what is necessary, is a physical-chemical explanation of the origins of these molecules that suggests the presence of suitable sources and sinks for free energy."

Proteinoid Microspheres

In trying to uncover the intermediate stages of abiogenesis mentioned by Bernal, Sidney W. Fox in the 1950s and 1960s studied the spontaneous formation of peptide structures under conditions that might plausibly have existed early in Earth's history. He demonstrated that amino acids could spontaneously form small chains called peptides. In one of his experiments, he allowed amino acids to dry out as if puddled in a warm, dry spot in prebiotic conditions. He found that, as they dried, the amino acids formed long, often cross-linked, thread-like, submicroscopic polypeptide molecules now named "proteinoid microspheres."

In another experiment using a similar method to set suitable conditions for life to form, Fox collected volcanic material from a cinder cone in Hawaii. He discovered that the temperature was over 100 °C (212 °F) just 4 inches (100 mm) beneath the surface of the cinder cone, and suggested that this might have been the environment in which life was created—molecules could have formed and then been washed through the loose volcanic ash and into the sea. He placed lumps of lava over amino acids derived from methane, ammonia and water, sterilized all materials, and baked the lava over the amino acids for a few hours in a glass oven. A brown, sticky substance formed over the surface and when the lava was drenched in sterilized water a thick, brown liquid leached out. It turned out that the amino acids had combined to form proteinoids, and the proteinoids had combined to form small globules that Fox called "microspheres." His proteinoids were not cells, although they formed clumps and chains reminiscent of cyanobacteria, but they contained no functional nucleic acids or any encoded information. Based upon such experiments, Colin S. Pittendrigh stated in December 1967 that "laboratories will be creating a living cell within ten years," a remark that reflected the typical contemporary levels of innocence of the complexity of cell structures.

Current Models

As of 2016, there is no single, generally accepted model for the origin of life. However, scientists have proposed several plausible theories, which share some common elements. While differing in the details, these theories are based on the framework laid out by Alexander Oparin (in 1924) and by J. B. S. Haldane (in 1925), who postulated the molecular or chemical evolution theory of life. According to them, the first molecules constituting the earliest cells "were synthesized under natural conditions by a slow process of molecular evolution, and these molecules then organized into the first molecular system with properties with biological order". Oparin and Haldane suggested that the atmosphere of the early Earth may have been chemically reducing in nature, composed primarily of methane (CH_4), ammonia (NH_3), water (H_2O), hydrogen sulfide (H_2S), carbon dioxide (CO_2) or carbon monoxide (CO), and phosphate (PO_4^{3-}), with molecular oxygen (O_2) and ozone (O_3) either rare or absent. According to later models, the atmosphere in the late Hadean period consisted largely of nitrogen (N_2) and carbon dioxide, with smaller amounts of carbon monoxide, hydrogen (H_2), and sulfur compounds; while it did lack molecular oxygen and ozone, it was not as chemically reducing as Oparin and Haldane supposed. In the atmosphere proposed by Oparin and Haldane, electrical activity can produce certain basic small molecules (monomers) of life, such as amino acids. The Miller–Urey experiment reported in 1953 demonstrated this.

Bernal coined the term *biopoiesis* in 1949 to refer to the origin of life. In 1967, he suggested that it occurred in three "stages":

1. the origin of biological monomers
2. the origin of biological polymers
3. the evolution from molecules to cells

Bernal suggested that evolution commenced between stages 1 and 2. The first stage is now fairly well understood, and the discovery of alkaline vents and the similarity with the "proton pump" found as the basis of biological life has begun to provide evidence about the second stage. Bernal regarded work on the third stage - discovering methods by which biological reactions were incorporated behind cell walls - as the most difficult. Modern work on the self-organizing capacities by which cell membranes self-assemble, and the work on micropores in various substrates is seen as a halfway house towards the development of independent free-living cells, and research into this continues.

The chemical processes that took place on the early Earth are called *chemical evolution*. Both Manfred Eigen and Sol Spiegelman demonstrated that evolution, including replication, variation, and natural selection, can occur in populations of molecules as well as in organisms. Spiegelman took advantage of natural selection to synthesize the Spiegelman Monster, which had a genome with just 218 nucleotide bases, having deconstructively evolved from a 4500-base bacterial RNA. Eigen built on Spiegelman's work and produced a similar system further degraded to just 48 or 54 nucleotides - the minimum required for the binding of the replication enzyme.

Following on from chemical evolution came the initiation of biological evolution, which led to the first cells. No one has yet synthesized a "protocell" using basic components with the necessary properties of life (the so-called "bottom-up-approach"). Without such a proof-of-principle, explanations have tended to focus on chemosynthesis. However, some researchers work in this field, notably Steen Rasmussen and Jack W. Szostak. Others have argued that a "top-down approach" is more feasible. One such approach, successfully attempted by Craig Venter and others at J. Craig Venter Institute, involves engineering existing prokaryotic cells with progressively fewer genes, attempting to discern at which point the most minimal requirements for life are reached.

Chemical Origin of Organic Molecules

The elements, except for hydrogen, ultimately derive from stellar nucleosynthesis. In October 2016, astronomers reported that the very basic chemical ingredients of life—the carbon-hydrogen molecule (CH, or methylidyne radical), the carbon-hydrogen positive ion (CH+) and the carbon ion (C+)—are the result, in large part, of ultraviolet light from stars, rather than in other ways, such as the result of turbulent events related to supernovae and young stars, as thought earlier. Complex molecules, including organic molecules, form naturally both in space and on planets. There are two possible sources of organic molecules on the early Earth:

1. Terrestrial origins – organic molecule synthesis driven by impact shocks or by other energy sources (such as UV light, redox coupling, or electrical discharges) (e.g., Miller's experiments)

2. Extraterrestrial origins – formation of organic molecules in interstellar dust clouds, which rain down on planets.

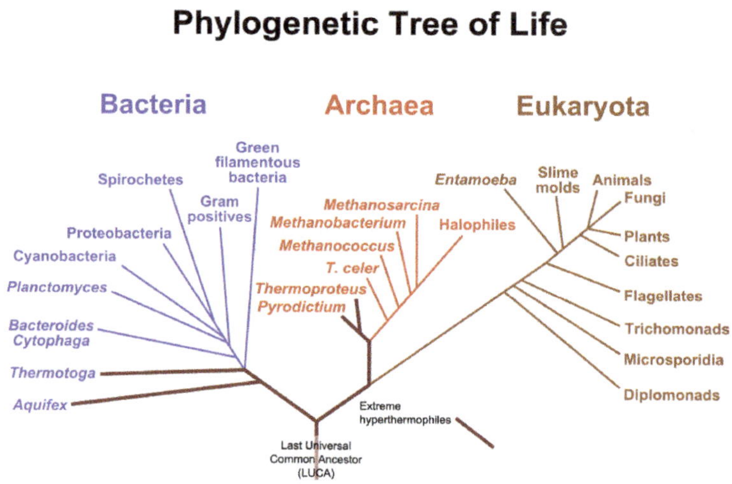

A cladogram demonstrating extreme hyperthermophiles at the base of the phylogenetic tree of life.

Based on recent computer model studies, the complex organic molecules necessary for life may have formed in the protoplanetary disk of dust grains surrounding the Sun before the formation of the Earth. According to the computer studies, this same process may also occur around other stars that acquire planets.

Estimates of the production of organics from these sources suggest that the Late Heavy Bombardment before 3.5 Ga within the early atmosphere made available quantities of organics comparable to those produced by terrestrial sources.

It has been estimated that the Late Heavy Bombardment may also have effectively sterilized the Earth's surface to a depth of tens of metres. If life evolved deeper than this, it would have also been shielded from the early high levels of ultraviolet radiation from the T Tauri stage of the Sun's evolution. Simulations of geothermically heated oceanic crust yield far more organics than those found in the Miller-Urey experiments. In the deep hydrothermal vents, Everett Shock has found "there is an enormous thermodynamic drive to form organic compounds, as seawater and hydrothermal fluids, which are far from equilibrium, mix and move towards a more stable state." Shock has found that the available energy is maximized at around 100 – 150 degrees Celsius, precisely the temperatures at which the hyperthermophilic bacteria and thermoacidophilic archaea have been found, at the base of the phylogenetic tree of life closest to the Last Universal Common Ancestor (LUCA).

Chemical Synthesis

While features of self-organization and self-replication are often considered the hallmark of living systems, there are many instances of abiotic molecules exhibiting such characteristics under proper conditions. Stan Palasek suggested based on a theoretical model that self-assembly of ribonucleic acid (RNA) molecules can occur spontaneously due to physical factors in hydrothermal vents. Virus self-assembly within host cells has implications for the study of the origin of life, as it lends further credence to the hypothesis that life could have started as self-assembling organic molecules.

Multiple sources of energy were available for chemical reactions on the early Earth. For example, heat (such as from geothermal processes) is a standard energy source for chemistry. Other examples include sunlight and electrical discharges (lightning), among others. Unfavourable reactions can also be driven by highly favourable ones, as in the case of iron-sulfur chemistry. For example, this was probably important for carbon fixation (the conversion of carbon from its inorganic form to an organic one).[note 1] Carbon fixation via iron-sulfur chemistry is highly favourable, and occurs at neutral pH and 100 °C (212 °F). Iron-sulfur surfaces, which are abundant near hydrothermal vents, are also capable of producing small amounts of amino acids and other biological metabolites.

Formamide produces all four ribonucleotides and other biological molecules when warmed in the presence of various terrestrial minerals. Formamide is ubiquitous in the Universe, produced by the reaction of water and hydrogen cyanide (HCN). It has several advantages as a biotic precursor, including the ability to easily become concentrated through the evaporation of water. Although HCN is poisonous, it only affects aerobic organisms (eukaryotes and aerobic bacteria), which did not yet exist. It can play roles in other chemical processes as well, such as the synthesis of the amino acid glycine.

In 1961, it was shown that the nucleic acid purine base adenine can be formed by heating aqueous ammonium cyanide solutions. Other pathways for synthesizing bases from inorganic materials were also reported. Leslie E. Orgel and colleagues have shown that freezing temperatures are advantageous for the synthesis of purines, due to the concentrating effect for key precursors such as hydrogen cyanide. Research by Stanley L. Miller and colleagues suggested that while adenine and guanine require freezing conditions for synthesis, cytosine and uracil may require boiling temperatures. Research by the Miller group notes the formation of seven different amino acids and 11 types of nucleobases in ice when ammonia and cyanide were left in a freezer from 1972 to 1997.

Other work demonstrated the formation of s-triazines (alternative nucleobases), pyrimidines (including cytosine and uracil), and adenine from urea solutions subjected to freeze-thaw cycles under a reductive atmosphere (with spark discharges as an energy source). The explanation given for the unusual speed of these reactions at such a low temperature is eutectic freezing. As an ice crystal forms, it stays pure: only molecules of water join the growing crystal, while impurities like salt or cyanide are excluded. These impurities become crowded in microscopic pockets of liquid within the ice, and this crowding causes the molecules to collide more often. Mechanistic exploration using quantum chemical methods provide a more detailed understanding of some of the chemical processes involved in chemical evolution, and a partial answer to the fundamental question of molecular biogenesis.

At the time of the Miller–Urey experiment, scientific consensus was that the early Earth had a reducing atmosphere with compounds relatively rich in hydrogen and poor in oxygen (e.g., CH_4 and NH_3 as opposed to CO_2 and nitrogen dioxide (NO_2)). However, current scientific consensus describes the primitive atmosphere as either weakly reducing or neutral. Such an atmosphere would diminish both the amount and variety of amino acids that could be produced, although studies that include iron and carbonate minerals (thought present in early oceans) in the experimental conditions have again produced a diverse array of amino acids. Other scientific research has focused on two other potential reducing environments: outer space and deep-sea thermal vents.

The spontaneous formation of complex polymers from abiotically generated monomers under the conditions posited by the "soup" theory is not at all a straightforward process. Besides the necessary basic organic monomers, compounds that would have prohibited the formation of polymers were also formed in high concentration during the Miller–Urey and Joan Oró experiments. The Miller–Urey experiment, for example, produces many substances that would react with the amino acids or terminate their coupling into peptide chains.

A research project completed in March 2015 by John D. Sutherland and others found that a network of reactions beginning with hydrogen cyanide and hydrogen sulfide, in streams of water irradiated by UV light, could produce the chemical components of proteins and lipids, as well as those of RNA, while not producing a wide range of other compounds. The researchers used the term "cyanosulfidic" to describe this network of reactions.

Autocatalysis

Autocatalysts are substances that catalyze the production of themselves and therefore are "molecular replicators." The simplest self-replicating chemical systems are autocatalytic, and typically contain three components: a product molecule and two precursor molecules. The product molecule joins together the precursor molecules, which in turn produce more product molecules from more precursor molecules. The product molecule catalyzes the reaction by providing a complementary template that binds to the precursors, thus bringing them together. Such systems have been demonstrated both in biological macromolecules and in small organic molecules. Systems that do not proceed by template mechanisms, such as the self-reproduction of micelles and vesicles, have also been observed.

It has been proposed that life initially arose as autocatalytic chemical networks. British ethologist Richard Dawkins wrote about autocatalysis as a potential explanation for the origin of life in his 2004 book *The Ancestor's Tale*. In his book, Dawkins cites experiments performed by Julius Rebek, Jr. and his colleagues in which they combined amino adenosine and pentafluorophenyl esters with the autocatalyst amino adenosine triacid ester (AATE). One product was a variant of AATE, which catalyzed the synthesis of themselves. This experiment demonstrated the possibility that autocatalysts could exhibit competition within a population of entities with heredity, which could be interpreted as a rudimentary form of natural selection.

In the early 1970s, Manfred Eigen and Peter Schuster examined the transient stages between the molecular chaos and a self-replicating hypercycle in a prebiotic soup. In a hypercycle, the information storing system (possibly RNA) produces an enzyme, which catalyzes the formation of another information system, in sequence until the product of the last aids in the formation of the first information system. Mathematically treated, hypercycles could create quasispecies, which through natural selection entered into a form of Darwinian evolution. A boost to hypercycle theory was the discovery of ribozymes capable of catalyzing their own chemical reactions. The hypercycle theory requires the existence of complex biochemicals, such as nucleotides, which do not form under the conditions proposed by the Miller–Urey experiment.

It has been shown that early error-prone translation machinery can be stable against an error catastrophe of the type that had been envisaged as problematical known as "Orgel's paradox" caused by catalytic activities that would be disruptive.

Homochirality

Homochirality refers to the geometric property of some materials that are composed of chiral units. Chiral refers to nonsuperimposable 3D forms that are mirror images of one another, as are left and right hands. Living organisms use molecules that have the same chirality ("handedness"): with almost no exceptions, amino acids are left-handed while nucleotides and sugars are right-handed. Chiral molecules can be synthesized, but in the absence of a chiral source or a chiral catalyst, they are formed in a 50/50 mixture of both enantiomers (called a racemic mixture). Known mechanisms for the production of non-racemic mixtures from racemic starting materials include: asymmetric physical laws, such as the electroweak interaction; asymmetric environments, such as those caused by circularly polarized light, quartz crystals, or the Earth's rotation; and statistical fluctuations during racemic synthesis.

Once established, chirality would be selected for. A small bias (enantiomeric excess) in the population can be amplified into a large one by asymmetric autocatalysis, such as in the Soai reaction. In asymmetric autocatalysis, the catalyst is a chiral molecule, which means that a chiral molecule is catalyzing its own production. An initial enantiomeric excess, such as can be produced by polarized light, then allows the more abundant enantiomer to outcompete the other.

Clark has suggested that homochirality may have started in outer space, as the studies of the amino acids on the Murchison meteorite showed that L-alanine is more than twice as frequent as its D form, and L-glutamic acid was more than three times prevalent than its D counterpart. Various chiral crystal surfaces can also act as sites for possible concentration and assembly of chiral monomer units into macromolecules. Compounds found on meteorites suggest that the chirality of life derives from abiogenic synthesis, since amino acids from meteorites show a left-handed bias, whereas sugars show a predominantly right-handed bias, the same as found in living organisms.

Self-enclosement, Reproduction, Duplication and the RNA World

Protocells

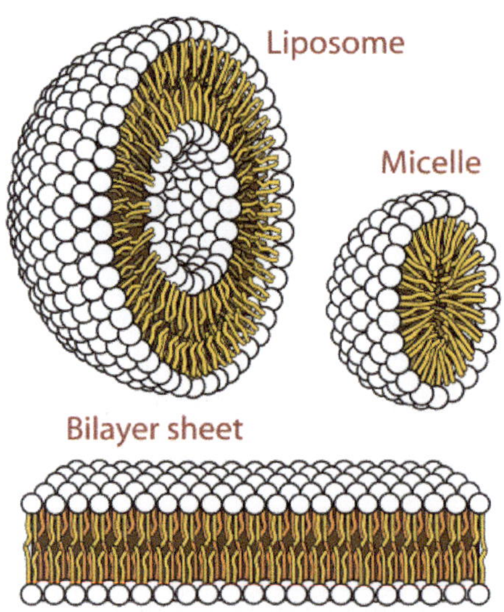

The three main structures phospholipids form spontaneously in solution: the liposome (a closed bilayer), the micelle and the bilayer.

A protocell is a self-organized, self-ordered, spherical collection of lipids proposed as a stepping-stone to the origin of life. A central question in evolution is how simple protocells first arose and differed in reproductive contribution to the following generation driving the evolution of life. Although a functional protocell has not yet been achieved in a laboratory setting, there are scientists who think the goal is well within reach.

Self-assembled vesicles are essential components of primitive cells. The second law of thermodynamics requires that the Universe move in a direction in which entropy increases, yet life is distinguished by its great degree of organization. Therefore, a boundary is needed to separate life processes from non-living matter. Researchers Irene A. Chen and Jack W. Szostak amongst others, suggest that simple physicochemical properties of elementary protocells can give rise to essential cellular behaviours, including primitive forms of differential reproduction competition and energy storage. Such cooperative interactions between the membrane and its encapsulated contents could greatly simplify the transition from simple replicating molecules to true cells. Furthermore, competition for membrane molecules would favour stabilized membranes, suggesting a selective advantage for the evolution of cross-linked fatty acids and even the phospholipids of today. Such micro-encapsulation would allow for metabolism within the membrane, the exchange of small molecules but the prevention of passage of large substances across it. The main advantages of encapsulation include the increased solubility of the contained cargo within the capsule and the storage of energy in the form of a electrochemical gradient.

A 2012 study led by Armen Y. Mulkidjanian of Germany's University of Osnabrück, suggests that inland pools of condensed and cooled geothermal vapour have the ideal characteristics for the origin of life. Scientists confirmed in 2002 that by adding a montmorillonite clay to a solution of fatty acid micelles (lipid spheres), the clay sped up the rate of vesicles formation 100-fold.

Another protocell model is the Jeewanu. First synthesized in 1963 from simple minerals and basic organics while exposed to sunlight, it is still reported to have some metabolic capabilities, the presence of semipermeable membrane, amino acids, phospholipids, carbohydrates and RNA-like molecules. However, the nature and properties of the Jeewanu remains to be clarified.

Electrostatic interactions induced by short, positively charged, hydrophobic peptides containing 7 amino acids in length or fewer, can attach RNA to a vesicle membrane, the basic cell membrane.

RNA World

The RNA world hypothesis describes an early Earth with self-replicating and catalytic RNA but no DNA or proteins. It is generally accepted that current life on Earth descends from an RNA world, although RNA-based life may not have been the first life to exist. This conclusion is drawn from many independent lines of evidence, such as the observations that RNA is central to the translation process and that small RNAs can catalyze all of the chemical groups and information transfers required for life. The structure of the ribosome has been called the "smoking gun," as it showed that the ribosome is a ribozyme, with a central core of RNA and no amino acid side chains within 18 angstroms of the active site where peptide bond formation is catalyzed. The concept of the RNA world was first proposed in 1962 by Alexander Rich, and the term was coined by Walter Gilbert in 1986.

Possible precursors for the evolution of protein synthesis include a mechanism to synthesize short peptide cofactors or form a mechanism for the duplication of RNA. It is likely that the ancestral ribosome was composed entirely of RNA, although some roles have since been taken over by proteins. Major remaining questions on this topic include identifying the selective force for the evolution of the ribosome and determining how the genetic code arose.

Eugene Koonin said, "Despite considerable experimental and theoretical effort, no compelling scenarios currently exist for the origin of replication and translation, the key processes that together comprise the core of biological systems and the apparent pre-requisite of biological evolution. The RNA World concept might offer the best chance for the resolution of this conundrum but so far cannot adequately account for the emergence of an efficient RNA replicase or the translation system. The MWO [Ed.: "many worlds in one"] version of the cosmological model of eternal inflation could suggest a way out of this conundrum because, in an infinite multiverse with a finite number of distinct macroscopic histories (each repeated an infinite number of times), emergence of even highly complex systems by chance is not just possible but inevitable."

Viral Origins and the RNA World

Recent evidence for a "virus first" hypothesis, which may support theories of the RNA world have been suggested in new research. One of the difficulties for the study of viral origins and evolution is their high rate of mutation; this is particularly the case in RNA retroviruses like HIV. A 2015 study compared protein fold structures across different branches of the tree of life, where researchers can reconstruct the evolutionary histories of the folds and of the organisms whose genomes code for those folds. They argue that protein folds are better markers of ancient events as their three-dimensional structures can be maintained even as the sequences that code for those begin to change. Thus, the viral protein repertoire retain traces of ancient evolutionary history that can be recovered using advanced bioinformatics approaches. Those researchers think that "the prolonged pressure of genome and particle size reduction eventually reduced virocells into modern viruses (identified by the complete loss of cellular makeup), meanwhile other coexisting cellular lineages diversified into modern cells. The data suggest that viruses originated from ancient cells that co-existed with the ancestors of modern cells. These ancient cells likely contained segmented RNA genomes.

RNA Synthesis and Replication

The RNA world hypothesis has spurred scientists to determine if RNA molecules could have spontaneously formed able to catalyze their own replication. Evidence suggests that the chemical conditions, including the presence of boron, molybdenum and oxygen needed for the initial production of RNA molecules, may have been better on the planet Mars than on the planet Earth. If so, life-suitable molecules originating on Mars, may have later migrated to Earth via meteor ejections.

A number of hypotheses of formation of RNA have been put forward. As of 1994, there are difficulties in the explanation of the abiotic synthesis of the nucleotides cytosine and uracil. Subsequent research has shown possible routes of synthesis; for example, formamide produces all four ribonucleotides and other biological molecules when warmed in the presence of various terrestrial minerals. Early cell membranes could have formed spontaneously from proteinoids, which are protein-like molecules produced when amino acid solutions are heated while in the correct concentration of aqueous solution. These are seen to form micro-spheres which are observed to behave similarly to membrane-enclosed compartments. Other possible means of producing more complicated organic molecules include chemical reactions that take place on clay substrates or on the surface of the mineral pyrite.

Factors supportive of an important role for RNA in early life include its ability to act both to store information and to catalyze chemical reactions (as a ribozyme); its many important roles as an intermediate in the expression of and maintenance of the genetic information (in the form of DNA) in modern organisms; and the ease of chemical synthesis of at least the components of the RNA molecule under the conditions that approximated the early Earth. Relatively short RNA molecules have been artificially produced in labs, which are capable of replication. Such replicase RNA, which functions as both code and catalyst provides its own template upon which copying can occur. Jack W. Szostak has shown that certain catalytic RNAs can join smaller RNA sequences together, creating the potential for self-replication. If these conditions were present, Darwinian natural selection would favour the proliferation of such autocatalytic sets, to which further functionalities could be added. Such autocatalytic systems of RNA capable of self-sustained replication have been identified. The RNA replication systems, which include two ribozymes that catalyze each other's synthesis, showed a doubling time of the product of about one hour, and were subject to natural selection under the conditions that existed in the experiment. In evolutionary competition experiments, this led to the emergence of new systems which replicated more efficiently. This was the first demonstration of evolutionary adaptation occurring in a molecular genetic system.

Depending on the specific definition used, life can be considered to have emerged when RNA chains began to express the basic conditions necessary for natural selection to operate as conceived by Darwin: heritability, variation of type, and differential reproductive output. The fitness of an RNA replicator (its per capita rate of increase) would likely be a function of its adaptive capacities that are intrinsic (in the sense that they were determined by the nucleotide sequence) and the availability of its resources. The three primary adaptive capacities may have been (1) the capacity to replicate with moderate fidelity, giving rise to both heritability while allowing variation of type, (2) the capacity to avoid decay, and (3) the capacity to acquire and process resources. These capacities would have been determined initially by the folded configurations of the RNA replicators that, in turn, would be encoded in their individual nucleotide sequences. Relative reproductive success, competition, between different replicators would have depended on the relative values of their adaptive capacities.

Pre-RNA World

It is possible that a different type of nucleic acid, such as PNA, TNA or GNA, was the first to emerge as a self-reproducing molecule, only later replaced by RNA. Larralde *et al.*, say that "the generally accepted prebiotic synthesis of ribose, the formose reaction, yields numerous sugars without any selectivity." and they conclude that their "results suggest that the backbone of the first genetic material could not have contained ribose or other sugars because of their instability." The ester linkage of ribose and phosphoric acid in RNA is known to be prone to hydrolysis.

Pyrimidine ribonucleosides and their respective nucleotides have been prebiotically synthesized by a sequence of reactions which by-pass the free sugars, and are assembled in a stepwise fashion by using nitrogenous or oxygenous chemistries. Sutherland has demonstrated high yielding routes to cytidine and uridine ribonucleotides built from small 2 and 3 carbon fragments such as glycolaldehyde, glyceraldehyde or glyceraldehyde-3-phosphate, cyanamide and cyanoacetylene. One of the steps in this sequence allows the isolation of enantiopure ribose aminooxazoline if the enantiomeric excess of glyceraldehyde is 60% or greater. This can be viewed as a prebiotic purification step, where the said compound spontaneously crystallized out from a mixture of the other pentose aminooxazolines. Ribose aminooxazoline can then react with cyanoacetylene in a mild and highly efficient manner to give the alpha cytidine ribonucleotide. Photoanomerization with UV light allows for inversion about the 1' anomeric centre to give the correct beta stereochemistry. In 2009 they showed that the same simple building blocks allow access, via phosphate controlled nucleobase elaboration, to 2',3'-cyclic pyrimidine nucleotides directly, which are known to be able to polymerize into RNA. This paper also highlights the possibility for the photo-sanitization of the pyrimidine-2',3'-cyclic phosphates.

Origin of Biological Metabolism

Metabolism-like reactions could have occurred naturally in early oceans, before the first organisms evolved. Metabolism may predate the origin of life and life may have evolved from the chemical conditions that prevailed in the world's earliest oceans. Reconstructions in laboratories show that some of these reactions can produce RNA, and some others resemble two essential reaction cascades of metabolism: glycolysis and the pentose phosphate pathway, that provide essential precursors for nucleic acids, amino acids and lipids. A study at the University of Dusseldorf created phylogenic trees based upon 6 million genes from bacteria and archaea, and identified 355 protein families that were probably present in LUCA. They were based upon an anaeobic metabolism fixing carbon dioxide and nitrogen. It suggests that LUCA evolved in an environment rich in hydrogen, carbon dioxide and iron. Following are some observed discoveries and related hypotheses.

Iron–sulfur World

In the 1980s, Günter Wächtershäuser, encouraged and supported by Karl R. Popper, postulated in his iron–sulfur world, a theory of the evolution of pre-biotic chemical pathways as the starting point in the evolution of life. It presents a consistent system of tracing today's biochemistry back to ancestral reactions that provide alternative pathways to the synthesis of organic building blocks from simple gaseous compounds.

In contrast to the classical Miller experiments, which depend on external sources of energy (such as simulated lightning or ultraviolet irradiation), "Wächtershäuser systems" come with a built-in source of energy, sulfides of iron (iron pyrite) and other minerals . The energy released from redox reactions of these metal sulfides is available for the synthesis of organic molecules. It is therefore hypothesized that such systems may be able to evolve into autocatalytic sets of self-replicating, metabolically active entities that predate the life forms known today. Experiments with such sulfides in an aqueous environment at 100 °C produced a relatively small yield of dipeptides (0.4% to 12.4%) and a smaller yield of tripeptides (0.10%) although under the same conditions, dipeptides were quickly broken down.

Several models reject the idea of the self-replication of a "naked-gene" but postulate the emergence of a primitive metabolism which could provide a safe environment for the later emergence of RNA replication. The centrality of the Krebs cycle (citric acid cycle) to energy production in aerobic organisms, and in drawing in carbon dioxide and hydrogen ions in biosynthesis of complex organic chemicals, suggests that it was one of the first parts of the metabolism to evolve. Somewhat in agreement with these notions, geochemist Michael Russell has proposed that "the purpose of life is to hydrogenate carbon dioxide" (as part of a "metabolism-first," rather than a "genetics-first," scenario). Physicist Jeremy England of MIT has proposed that thermodynamically, life was bound to eventually arrive, as based on established physics, he mathematically indicates "...that when a group of atoms is driven by an external source of energy (like the sun or chemical fuel) and surrounded by a heat bath (like the ocean or atmosphere), it will often gradually restructure itself in order to dissipate increasingly more energy. This could mean that under certain conditions, matter inexorably acquires the key physical attribute associated with life."

One of the earliest incarnations of this idea was put forward in 1924 with Oparin's notion of primitive self-replicating vesicles which predated the discovery of the structure of DNA. Variants in the 1980s and 1990s include Wächtershäuser's iron–sulfur world theory and models introduced by Christian de Duve based on the chemistry of thioesters. More abstract and theoretical arguments for the plausibility of the emergence of metabolism without the presence of genes include a mathematical model introduced by Freeman Dyson in the early 1980s and Stuart Kauffman's notion of collectively autocatalytic sets, discussed later in that decade.

Orgel summarized his analysis of the proposal by stating, "There is at present no reason to expect that multistep cycles such as the reductive citric acid cycle will self-organize on the surface of FeS/FeS_2 or some other mineral." It is possible that another type of metabolic pathway was used at the beginning of life. For example, instead of the reductive citric acid cycle, the "open" acetyl-CoA pathway (another one of the five recognized ways of carbon dioxide fixation in nature today) would be compatible with the idea of self-organization on a metal sulfide surface. The key enzyme of this pathway, carbon monoxide dehydrogenase/acetyl-CoA synthase harbours mixed nickel-iron-sulfur clusters in its reaction centres and catalyzes the formation of acetyl-CoA (which may be regarded as a modern form of acetyl-thiol) in a single step. There are increasing concerns, however, that prebiotic thiolated (i.e.Thioacetic acid) and Thioester compounds are thermodynamically and kinetically unfavourable to accumulate in presumed prebiotic conditions (i.e. Hydrothermal vents).

Zn-world Hypothesis

The Zn-world (zinc world) theory of Armen Y. Mulkidjanian is an extension of Wächtershäuser's pyrite hypothesis. Wächtershäuser based his theory of the initial chemical processes leading to informational molecules (i.e., RNA, peptides) on a regular mesh of electric charges at the surface of pyrite that may have made the primeval polymerization thermodynamically more favourable by attracting reactants and arranging them appropriately relative to each other. The Zn-world theory specifies and differentiates further. Hydrothermal fluids rich in H_2S interacting with cold primordial ocean (or Darwin's "warm little pond") water leads to the precipitation of metal sulfide particles. Oceanic vent systems and other hydrothermal systems have a zonal structure reflected in ancient volcanogenic massive sulfide deposits (VMS) of hydrothermal origin. They reach many kilometres in diameter and date back to the Archean Eon. Most abundant are pyrite (FeS_2), chalcopyrite ($CuFeS_2$), and sphalerite (ZnS), with additions of galena (PbS) and alabandite (MnS). ZnS and MnS have a unique ability to store radiation energy, e.g., provided by UV light. Since during the relevant time window of the origins of replicating molecules the primordial atmospheric pressure was high enough (>100 bar, about 100 atmospheres) to precipitate near the Earth's surface and UV irradiation was 10 to 100 times more intense than now, the unique photosynthetic properties mediated by ZnS provided just the right energy conditions to energize the synthesis of informational and metabolic molecules and the selection of photostable nucleobases.

The Zn-world theory has been further filled out with experimental and theoretical evidence for the ionic constitution of the interior of the first proto-cells before archaea, bacteria and proto-eukaryotes evolved. Archibald Macallum noted the resemblance of organism fluids such as blood, and lymph to seawater; however, the inorganic composition of all cells differ from that of modern seawater, which led Mulkidjanian and colleagues to reconstruct the "hatcheries" of the first cells combining geochemical analysis with phylogenomic scrutiny of the inorganic ion requirements of universal components of modern cells. The authors conclude that ubiquitous, and by inference primordial, proteins and functional systems show affinity to and functional requirement for K^+, Zn^{2+}, Mn^{2+}, and phosphate. Geochemical reconstruction shows that the ionic composition conducive to the origin of cells could not have existed in what we today call marine settings but is compatible with emissions of vapour-dominated zones of what we today call inland geothermal systems. Under the oxygen depleted, CO_2-dominated primordial atmosphere, the chemistry of water condensates and exhalations near geothermal fields would resemble the internal milieu of modern cells. Therefore, the precellular stages of evolution may have taken place in shallow "Darwin ponds" lined with porous silicate minerals mixed with metal sulfides and enriched in K^+, Zn^{2+}, and phosphorus compounds.

Deep Sea Vent Hypothesis

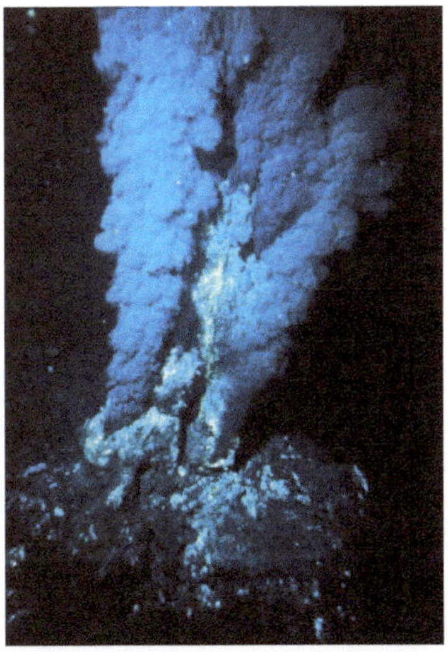

Deep-sea hydrothermal vent or 'black smoker'

The deep sea vent, or alkaline hydrothermal vent, theory for the origin of life on Earth posits that life may have begun at submarine hydrothermal vents, William Martin and Michael Russell have suggested "that life evolved in structured iron monosulphide precipitates in a seepage site hydrothermal mound at a redox, pH and temperature gradient between sulphide-rich hydrothermal fluid and iron(II)-containing waters of the Hadean ocean floor. The naturally arising, three-dimensional compartmentation observed within fossilized seepage-site metal sulphide precipitates indicates that these inorganic compartments were the precursors of cell walls and membranes found in free-living prokaryotes. The known capability of FeS and NiS to catalyze the synthesis of the acetyl-methylsulphide from carbon monoxide and methylsulphide, constituents of hydrothermal fluid, indicates that pre-biotic syntheses occurred at the inner surfaces of these metal-sulphide-walled compartments,..." These form where hydrogen-rich fluids emerge from below the sea floor, as a result of serpentinization of ultra-mafic olivine with seawater and a pH interface with carbon dioxide-rich ocean water. The vents form a sustained chemical energy source derived from redox reactions, in which electron donors, such as molecular hydrogen, react with electron acceptors, such as carbon dioxide. These are highly exothermic reactions.

Michael Russell demonstrated that alkaline vents created an abiogenic proton motive force (PMF) chemiosmotic gradient, in which conditions are ideal for an abiogenic hatchery for life. Their microscopic compartments "provide a natural means of concentrating organic molecules," composed of iron-sulfur minerals such as mackinawite, endowed these mineral cells with the catalytic properties envisaged by Wächtershäuser. This movement of ions across the membrane depends on a combination of two factors:

1. Diffusion force caused by concentration gradient—all particles including ions tend to diffuse from higher concentration to lower.
2. Electrostatic force caused by electrical potential gradient—cations like protons H^+ tend to diffuse down the electrical potential, anions in the opposite direction.

These two gradients taken together can be expressed as an electrochemical gradient, providing energy for abiogenic synthesis. The proton motive force can be described as the measure of the potential energy stored as a combination of proton and voltage gradients across a membrane (differences in proton concentration and electrical potential).

White smokers emitting liquid carbon dioxide (CO_2) at the *Champagne* vent, Marianas Trench Marine National Monument

Jack W. Szostak suggested that geothermal activity provides greater opportunities for the origination of life in open lakes where there is a buildup of minerals. In 2010, based on spectral analysis of sea and hot mineral water, Ignat Ignatov and Oleg Mosin demonstrated that life may have predominantly originated in hot mineral water. The hot mineral water that contains bicarbonate and calcium ions has the most optimal range. This case is similar to the origin of life in hydrothermal vents, but with bicarbonate and calcium ions in hot water. This water has a pH of 9–11 and is possible to have the reactions in seawater. According to Melvin Calvin, certain reactions of condensation-dehydration of amino acids and nucleotides in individual blocks of peptides and nucleic acids can take place in the primary hydrosphere with pH 9-11 at a later evolutionary stage. Some of these compounds like hydrocyanic acid (HCN) have been proven in the experiments of Miller. This is the environment in which the stromatolites have been created. David Ward of Montana State University described the formation of stromatolites in hot mineral water at the Yellowstone National Park. Stromatolites survive in hot mineral water and in proximity to areas with volcanic activity. Processes have evolved in the sea near geysers of hot mineral water. In 2011, Tadashi Sugawara from the University of Tokyo created a protocell in hot water.

Experimental research and computer modelling suggest that the surfaces of mineral particles inside hydrothermal vents have catalytic properties similar to those of enzymes and are able to create simple organic molecules, such as methanol (CH_3OH) and formic, acetic and pyruvic acid out of the dissolved CO_2 in the water.

The research reported above by William F. Martin in July 2016 supports the thesis that life arose at hydrothermal vents, that spontaneous chemistry in the Earth's crust driven by rock–water interactions at disequilibrium thermodynamically underpinned life's origin and that the founding lineages of the archaea and bacteria were H2-dependent autotrophs that used CO2 as their terminal acceptor in energy metabolism. Martin suggests, based upon this evidence that LUCA "may have depended heavily on the geothermal energy of the vent to survive".

Thermosynthesis

Today's bioenergetic process of fermentation is carried out by either the aforementioned citric acid cycle or the Acetyl-CoA pathway, both of which have been connected to the primordial Iron–sulfur world. In a different approach, the thermosynthesis hypothesis considers the bioenergetic process of chemiosmosis, which plays an essential role in cellular respiration and photosynthesis, more basal than fermentation: the ATP synthase enzyme, which sustains chemiosmosis, is proposed as the currently extant enzyme most closely related to the first metabolic process.

First, life needed an energy source to bring about the condensation reaction that yielded the peptide bonds of proteins and the phosphodiester bonds of RNA. In a generalization and thermal variation of the binding change mechanism of today's ATP synthase, the "first protein" would have bound substrates (peptides, phosphate, nucleosides, RNA 'monomers') and condensed them to a reaction product that remained bound until after a temperature change it was released by thermal unfolding.

The energy source under the thermosynthesis hypothesis was thermal cycling, the result of suspension of protocells in a convection current, as is plausible in a volcanic hot spring; the convection accounts for the self-organization and dissipative structure required in any origin of life model. The still ubiquitous role of thermal cycling in germination and cell division is considered a relic of primordial thermosynthesis.

By phosphorylating cell membrane lipids, this "first protein" gave a selective advantage to the lipid protocell that contained the protein. This protein also synthesized a library of many proteins, of which only a minute fraction had thermosynthesis capabilities. As proposed by Dyson, it propagated functionally: it made daughters with similar capabilities, but it did not copy itself. Functioning daughters consisted of different amino acid sequences.

Whereas the Iron–sulfur world identifies a circular pathway as the most simple, the thermosynthesis hypothesis does not even invoke a pathway: ATP synthase's binding change mechanism resembles a physical adsorption process that yields free energy, rather than a regular enzyme's mechanism, which decreases the free energy. It has been claimed that the emergence of cyclic systems of protein catalysts is implausible.

Other Models of Abiogenesis

Clay Hypothesis

Montmorillonite, an abundant clay, is a catalyst for the polymerization of RNA and for the formation of membranes from lipids. A model for the origin of life using clay was forwarded by Alexander Graham Cairns-Smith in 1985 and explored as a plausible mechanism by several scientists. The clay hypothesis postulates that complex organic molecules arose gradually on a pre-existing, non-organic replication surfaces of silicate crystals in solution.

At the Rensselaer Polytechnic Institute, James P. Ferris' studies have also confirmed that clay minerals of montmorillonite catalyze the formation of RNA in aqueous solution, by joining nucleotides to form longer chains.

In 2007, Bart Kahr from the University of Washington and colleagues reported their experiments that tested the idea that crystals can act as a source of transferable information, using crystals of potassium hydrogen phthalate. "Mother" crystals with imperfections were cleaved and used as seeds to grow "daughter" crystals from solution. They then examined the distribution of imperfections in the new crystals and found that the imperfections in the mother crystals were reproduced in the daughters, but the daughter crystals also had many additional imperfections. For gene-like behaviour to be observed, the quantity of inheritance of these imperfections should have exceeded that of the mutations in the successive generations, but it did not. Thus Kahr concluded that the crystals "were not faithful enough to store and transfer information from one generation to the next."

Gold's "Deep-hot Biosphere" Model

In the 1970s, Thomas Gold proposed the theory that life first developed not on the surface of the Earth, but several kilometres below the surface. It is claimed that discovery of microbial life below the surface of another body in our Solar System would lend significant credence to this theory. Thomas Gold also asserted that a trickle of food from a deep, unreachable, source is needed for survival because life arising in a puddle of organic material is likely to consume all of its food and become extinct. Gold's theory is that the flow of such food is due to out-gassing of primordial methane from the Earth's mantle; more conventional explanations of the food supply of deep microbes (away from sedimentary carbon compounds) is that the organisms subsist on hydrogen released by an interaction between water and (reduced) iron compounds in rocks.

Panspermia

Panspermia is the hypothesis that life exists throughout the Universe, distributed by meteoroids, asteroids, comets, planetoids, and, also, by spacecraft in the form of unintended contamination by microorganisms.

Panspermia hypothesis does not attempt to explain how life first originated, but merely shifts it to another planet or a comet. The advantage of an extraterrestrial origin of primitive life is that life is not required to have formed on each planet it occurs on, but rather in a single location, and then spread about the galaxy to other star systems via cometary and/or meteorite impact. Evidence to support the hypothesis is scant, but it finds support in studies of Martian meteorites found in Antarctica and in studies of extremophile microbes' survival in outer space tests.

Extraterrestrial Organic Molecules

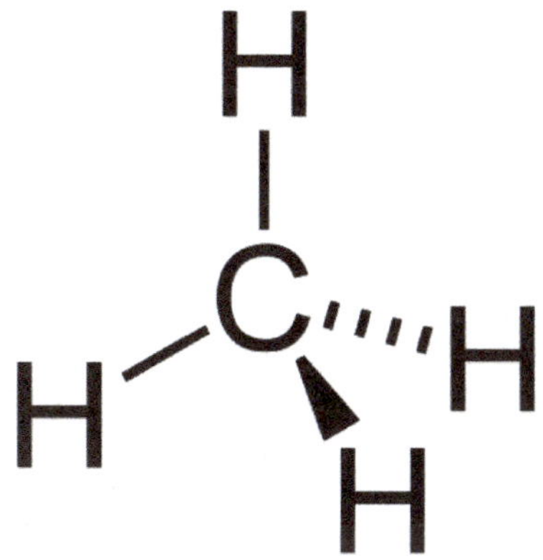

Methane is one of the simplest organic compounds

An organic compound is any member of a large class of gaseous, liquid, or solid chemicals whose molecules contain carbon. Carbon is the fourth most abundant element in the Universe by mass after hydrogen, helium, and oxygen. Carbon is abundant in the Sun, stars, comets, and in the atmospheres of most planets. Organic compounds are relatively common in space, formed by "factories of complex molecular synthesis" which occur in molecular clouds and circumstellar envelopes, and chemically evolve after reactions are initiated mostly by ionizing radiation. Based on computer model studies, the complex organic molecules necessary for life may have formed on dust grains in the protoplanetary disk surrounding the Sun before the formation of the Earth. According to the computer studies, this same process may also occur around other stars that acquire planets.

Observations suggest that the majority of organic compounds introduced on Earth by interstellar dust particles are considered principal agents in the formation of complex molecules, thanks to their peculiar surface-catalytic activities. Studies reported in 2008, based on $^{12}C/^{13}C$ isotopic ratios of organic compounds found in the Murchison meteorite, suggested that the RNA component uracil and related molecules, including xanthine, were formed extraterrestrially. On 8 August 2011, a report based on NASA studies of meteorites found on Earth was published suggesting DNA components (adenine, guanine and related organic molecules) were made in outer space. Scientists also found that the cosmic dust permeating the Universe contains complex organics ("amorphous organic solids with a mixed aromatic–aliphatic structure") that could be created naturally, and rapidly, by stars. Sun Kwok of The University of Hong Kong suggested that these compounds may have been related to the development of life on Earth said that "If this is the case, life on Earth may have had an easier time getting started as these organics can serve as basic ingredients for life."

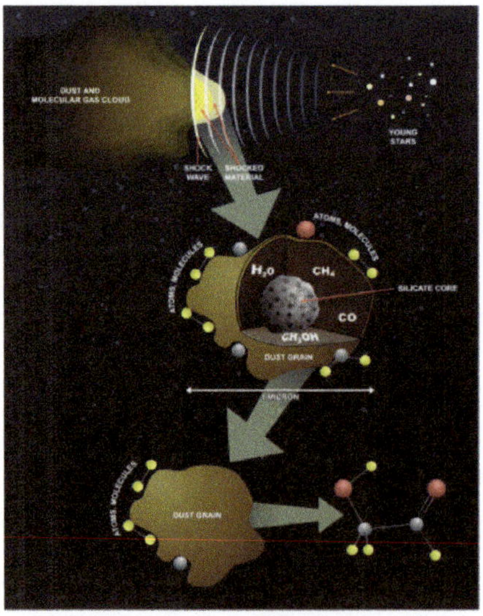

Formation of glycolaldehyde in stardust

Glycolaldehyde, the first example of an interstellar sugar molecule, was detected in the star-forming region near the centre of our galaxy. It was discovered in 2000 by Jes Jørgensen and Jan M. Hollis. In 2012, Jørgensen's team reported the detection of glycolaldehyde in a distant star system. The molecule was found around the protostellar binary IRAS 16293-2422 400 light years from Earth. Glycolaldehyde is needed to form RNA, which is similar in function to DNA. These findings suggest that complex organic molecules may form in stellar systems prior to the formation of planets, eventually arriving on young planets early in their formation. Because sugars are associated with both metabolism and the genetic code, two of the most basic aspects of life, it is thought the discovery of extraterrestrial sugar increases the likelihood that life may exist elsewhere in our galaxy.

NASA announced in 2009 that scientists had identified another fundamental chemical building block of life in a comet for the first time, glycine, an amino acid, which was detected in material ejected from comet Wild 2 in 2004 and grabbed by NASA's *Stardust* probe. Glycine has been detected in meteorites before. Carl Pilcher, who leads the NASA Astrobiology Institute commented that "The discovery of glycine in a comet supports the idea that the fundamental building blocks of life are prevalent in space, and strengthens the argument that life in the Universe may be common rather than rare." Comets are encrusted with outer layers of dark material, thought to be a tar-like substance composed of complex organic material formed from simple carbon compounds after reactions initiated mostly by ionizing radiation. It is possible that a rain of material from comets could have brought significant quantities of such complex organic molecules to Earth. Amino acids which were formed extraterrestrially may also have arrived on Earth via comets. It is estimated that during the Late Heavy Bombardment, meteorites may have delivered up to five million tons of organic prebiotic elements to Earth per year.

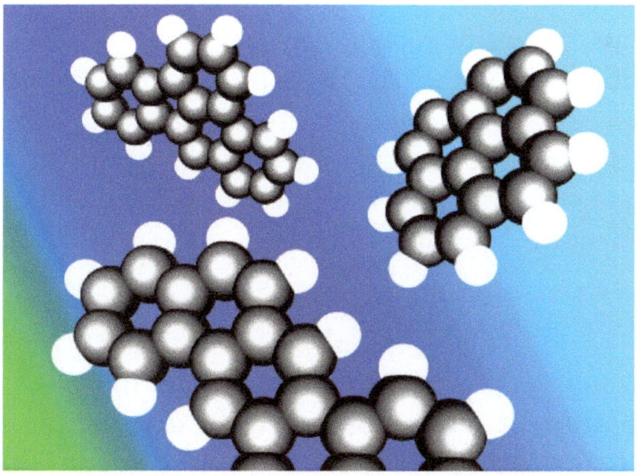

An illustration of typical polycyclic aromatic hydrocarbons. Clockwise from top left: benz(e)acephenanthrylene, pyrene and dibenz(ah)anthracene.

Polycyclic aromatic hydrocarbons (PAH) are the most common and abundant of the known polyatomic molecules in the observable universe, and are considered a likely constituent of the primordial sea. In 2010, PAHs, along with fullerenes (or "buckyballs"), have been detected in nebulae.

In March 2015, NASA scientists reported that, for the first time, complex DNA and RNA organic compounds of life, including uracil, cytosine and thymine, have been formed in the laboratory under outer space conditions, using starting chemicals, such as pyrimidine, found in meteorites. Pyrimidine, like PAHs, the most carbon-rich chemical found in the Universe, may have been formed in red giant stars or in interstellar dust and gas clouds.

Lipid World

The lipid world theory postulates that the first self-replicating object was lipid-like. It is known that phospholipids form lipid bilayers in water while under agitation—the same structure as in cell membranes. These molecules were not present on early Earth, but other amphiphilic long-chain molecules also form membranes. Furthermore, these bodies may expand (by insertion of additional lipids), and under excessive expansion may undergo spontaneous splitting which preserves the same size and composition of lipids in the two progenies. The main idea in this theory is that the molecular composition of the lipid bodies is the preliminary way for information storage, and evolution led to the appearance of polymer entities such as RNA or DNA that may store information favourably. Studies on vesicles from potentially prebiotic amphiphiles have so far been limited to systems containing one or two types of amphiphiles. This in contrast to the output of simulated prebiotic chemical reactions, which typically produce very heterogeneous mixtures of compounds. Within the hypothesis of a lipid bilayer membrane composed of a mixture of various distinct amphiphilic compounds there is the opportunity of a huge number of theoretically possible combinations in the arrangements of these amphiphiles in the membrane. Among all these potential combinations, a specific local arrangement of the membrane would have favoured the constitution of a hypercycle, actually a positive feedback composed of two mutual catalysts represented by a membrane site and a specific compound trapped in the vesicle. Such site/compound pairs are transmissible to the daughter vesicles leading to the emergence of distinct lineages of vesicles which would have allowed Darwinian natural selection.

Polyphosphates

A problem in most scenarios of abiogenesis is that the thermodynamic equilibrium of amino acid versus peptides is in the direction of separate amino acids. What has been missing is some force that drives polymerization. The resolution of this problem may well be in the properties of polyphosphates. Polyphosphates are formed by polymerization of ordinary monophosphate ions PO_4^{-3}. Several mechanisms for such polymerization have been suggested. Polyphosphates cause polymerization of amino acids into peptides. They are also logical precursors in the synthesis of such key biochemical compounds as adenosine triphosphate (ATP). A key issue seems to be that calcium reacts with soluble phosphate to form insoluble calcium phosphate (apatite), so some plausible mechanism must be found to keep calcium ions from causing precipitation of phosphate. There has been much work on this topic over the years, but an interesting new idea is that meteorites may have introduced reactive phosphorus species on the early Earth.

PAH World Hypothesis

Polycyclic aromatic hydrocarbons (PAH) are known to be abundant in the Universe, including in the interstellar medium, in comets, and in meteorites, and are some of the most complex molecules so far found in space.

Other sources of complex molecules have been postulated, including extraterrestrial stellar or interstellar origin. For example, from spectral analyses, organic molecules are known to be present in comets and meteorites. In 2004, a team detected traces of PAHs in a nebula. In 2010, another team also detected PAHs, along with fullerenes, in nebulae. The use of PAHs has also been proposed as a precursor to the RNA world in the PAH world hypothesis. The Spitzer Space Telescope has detected a star, HH 46-IR, which is forming by a process similar to that by which the Sun formed. In the disk of material surrounding the star, there is a very large range of molecules, including cyanide compounds, hydrocarbons, and carbon monoxide. In September 2012, NASA scientists reported that PAHs, subjected to interstellar medium conditions, are transformed, through hydrogenation, oxygenation and hydroxylation, to more complex organics—"a step along the path toward amino acids and nucleotides, the raw materials of proteins and DNA, respectively." Further, as a result of these transformations, the PAHs lose their spectroscopic signature which could be one of the reasons "for the lack of PAH detection in interstellar ice grains, particularly the outer regions of cold, dense clouds or the upper molecular layers of protoplanetary disks."

NASA maintains a database for tracking PAHs in the Universe. More than 20% of the carbon in the Universe may be associated with PAHs, possible starting materials for the formation of life. PAHs seem to have been formed shortly after the Big Bang, are widespread throughout the Universe, and are associated with new stars and exoplanets.

Radioactive Beach Hypothesis

Zachary Adam claims that tidal processes that occurred during a time when the Moon was much closer may have concentrated grains of uranium and other radioactive elements at the high-water mark on primordial beaches, where they may have been responsible for generating life's building blocks. According to computer models reported in *Astrobiology*, a deposit of such radioactive materials could show the same self-sustaining nuclear reaction as that found in the Oklo uranium ore seam in Gabon. Such radioactive beach sand might have provided sufficient energy to generate organic molecules, such as amino acids and sugars from acetonitrile in water. Radioactive monazite material also has released soluble phosphate into the regions between sand-grains, making it biologically "accessible." Thus amino acids, sugars, and soluble phosphates might have been produced simultaneously, according to Adam. Radioactive actinides, left behind in some concentration by the reaction, might have formed part of organometallic complexes. These complexes could have been important early catalysts to living processes.

John Parnell has suggested that such a process could provide part of the "crucible of life" in the early stages of any early wet rocky planet, so long as the planet is large enough to have generated a system of plate tectonics which brings radioactive minerals to the surface. As the early Earth is thought to have had many smaller plates, it might have provided a suitable environment for such processes.

Thermodynamic Dissipation

Karo Michaelian from the National Autonomous University of Mexico (UNAM) points out that any model for the origin of life must take into account the fact that life is an irreversible thermodynamic process and, like all irreversible processes, its origin and persistence as a "self-organized" system is due to its dissipation an imposed generalized chemical potential, i.e., the production of entropy. That is, entropy production is not incidental to the process of life, but rather the fundamental reason for its existence. Present day life augments the entropy production of Earth in its solar environment by dissipating ultraviolet and visible photons into heat through organic pigments in water. This heat then catalyzes a host of secondary dissipative processes such as the water cycle, ocean and wind currents, hurricanes, etc. Michaelian argues that if the thermodynamic function of life today is to produce entropy through photon dissipation, then this probably was its function at its very beginnings. It turns out that both RNA and DNA when in water solution are very strong absorbers and extremely rapid dissipaters of UV light within the 230–290 nm wavelength region, which is a part of the Sun's spectrum that could have penetrated the prebiotic atmosphere. The amount of ultraviolet (UV-C) light reaching the Earth's surface within this spectral range in the Archean could have been on the order of 4 W/m^2, or some 31 orders of magnitude greater than it is today at 260 nm where RNA and DNA absorb most strongly. In fact, not only RNA and DNA, but many fundamental molecules of life (those common to all three domains of life, archea, bacteria, and eucaryote) are also pigments that absorb in the UV-C, and many of these also have a chemical affinity to RNA and DNA. Nucleic acids may thus have acted as acceptor molecules to the UV-C photon excited antenna pigment donor molecules by providing an ultrafast channel for dissipation. Michaelian has shown that there would have existed a non-linear, non-equilibrium thermodynamic imperative to the abiogenic UV-C photochemical synthesis and proliferation of these pigments over the entire Earth surface if they augmented the solar photon dissipation rate.

A simple mechanism to explain enzyme-less replication of RNA and DNA can be given within the same dissipative thermodynamic framework by assuming that life arose when the temperature of the primitive seas had cooled to somewhat below the denaturing temperature of RNA or DNA. The ratio of $^{18}O/^{16}O$ found in cherts of the Barberton greenstone belt of South Africa indicates that the Earth's surface temperature was around 80 °C at 3.8 Ga, falling to 70±15 °C about 3.5 to 3.2 Ga, suggestively close to RNA or DNA denaturing (uncoiling and separation) temperatures. During the night, the surface water temperature would drop below the denaturing temperature and single strand RNA/DNA could act as extension template for the formation of double strand RNA/DNA. During the daylight hours, RNA and DNA would absorb UV-C light and convert this directly into heat at the ocean surface, thereby raising the local temperature enough to allow for denaturing of RNA and DNA. Direct experimental evidence for the denaturing of DNA through UV-C light dissipation has now been obtained.

The copying process would have been repeated with each diurnal cycle. Such an ultraviolet and temperature assisted RNA/DNA reproduction (UVTAR) bears similarity to polymerase chain reaction (PCR), a routine laboratory procedure employed to multiply DNA segments. Since denaturation would be most probable in the late afternoon when the Archean sea surface temperature would be highest, and since late afternoon submarine sunlight is somewhat circularly polarized, the homochirality of the organic molecules of life can also be explained within the proposed thermodynamic framework.

The fact that the aromatic amino acids have been shown to have chemical affinity to their codons, or anti-codons, and that they also absorb strongly in the UV-C, suggests that they might have originally acted as antenna pigments to increase dissipation and to provide more local heat for UVTAR replication of RNA and DNA as the sea surface temperature cooled. The accumulation of information, e.g., coding for the aromatic amino acids, in RNA or DNA would thus be related to reproductive success under this mechanism. Michaelian suggests that the traditional origin of life research, that expects to describe the emergence of life without overwhelming reference to entropy production through dissipation, is erroneous and that imposed environmental potentials, such as the solar photon flux, and the dissipation of this flux, must be considered to understand the emergence, proliferation, and evolution of life.

Multiple Genesis

Different forms of life with variable origin processes may have appeared quasi-simultaneously in the early history of Earth. The other forms may be extinct (having left distinctive fossils through their different biochemistry—e.g., hypothetical types of biochemistry). It has been proposed that:

The first organisms were self-replicating iron-rich clays which fixed carbon dioxide into oxalic and other dicarboxylic acids. This system of replicating clays and their metabolic phenotype then evolved into the sulfide rich region of the hotspring acquiring the ability to fix nitrogen. Finally phosphate was incorporated into the evolving system which allowed the synthesis of nucleotides and phospholipids. If biosynthesis recapitulates biopoiesis, then the synthesis of amino acids preceded the synthesis of the purine and pyrimidine bases. Furthermore the polymerization of the amino acid thioesters into polypeptides preceded the directed polymerization of amino acid esters by polynucleotides.

Fluctuating Hydrothermal Pools on Volcanic Islands or Proto-continents

Armid Mulkidjanian and co-authors think that the marine environments did not provide the ionic balance and composition universally found in cells, as well as of ions required by essential proteins and ribozymes found in virtually all living organisms, especially with respect to K^+/Na^+ ratio, Mn^{2+}, Zn^{2+} and phosphate concentrations. The only known environments that mimic the needed conditions on Earth are found in terrestrial hydrothermal pools fed by steam vents. Additionally, mineral deposits in these environments under an anoxic atmosphere would have suitable pH (as opposed to current pools in an oxygenated atmosphere), contain precipitates of sulfide minerals that block harmful UV radiation, have wetting/drying cycles that concentrate substrate solutions to concentrations amenable to spontaneous formation of polymers of nucleic acids, and a continual supply of abiotically generated organic molecules, both by chemical reactions in the hydrothermal environment, as well as by exposure to UV light during transport from vents to adjacent pools. Their hypothesized pre-biotic environments are similar to the deep-oceanic vent environments most commonly hypothesized, but add additional components that help explain peculiarities found in reconstructions of the Last Universal Common Ancestor (LUCA) of all living organisms.

Bruce Damer and David Deamer have come to the conclusion that cell membranes cannot be formed in salty seawater, and must therefore have originated in freshwater. Before the continents formed, the only dry land on Earth would be volcanic islands, where rainwater would form ponds where lipids could form the first stages towards cell membranes. These predecessors of true cells are assumed to have behaved more like a superorganism rather than individual structures, where the porous membranes would house molecules which would leak out and enter other protocells. Only when true cells had evolved would they gradually adapt to saltier environments and enter the ocean.

Information Theory

A theory that speaks to the origin of life on Earth and other rocky planets posits life as an information system in which information content grows because of selection. Life must start with minimum possible information, or minimum possible departure from thermodynamic equilibrium, and it requires thermodynamically free energy accessible by means of its information content. The most benign circumstances, minimum entropy variations with abundant free energy, suggest the pore space in the first few kilometres of the surface. Free energy is derived from the condensed products of the chemical reactions taking place in the cooling nebula.

Human Ecology

Part of the built environment – suburban tract housing in Colorado Springs, Colorado

Human Ecology is an interdisciplinary and transdisciplinary study of the relationship between humans and their natural, social, and built environments. The philosophy and study of human ecology has a diffuse history with advancements in geography, sociology, psychology, anthropology, zoology, epidemiology, public health, home economics, and natural ecology, among others.

Historical Development

The roots of ecology as a broader discipline can be traced to the Greeks and a lengthy list of developments in natural history science. Ecology also has notably developed in other cultures. Traditional knowledge, as it is called, includes the human propensity for intuitive knowledge, intelligent relations, understanding, and for passing on information about the natural world and the human experience. The term ecology was coined by Ernst Haeckel in 1866 and defined by direct reference to *the economy of nature.*

Like other contemporary researchers of his time, Haeckel adopted his terminology from Carl Linnaeus where human ecological connections were more evident. In his 1749 publication, *Specimen academicum de oeconomia naturae*, Linnaeus developed a science that included the *economy and polis of nature.* Polis stems from its Greek roots for a political community (originally based on the city-states), sharing its roots with the word police in reference to the promotion of growth and maintenance of good social order in a community. Linnaeus was also the first to write about the close affinity between humans and primates. Linnaeus presented early ideas found in modern aspects to human ecology, including the balance of nature while highlighting the importance of ecological functions (ecosystem services or natural capital in modern terms): "In exchange for performing its function satisfactorily, nature provided a species with the necessaries of life" The work of Linnaeus influenced Charles Darwin and other scientists of his time who used Linnaeus' terminology (i.e., the *economy and polis of nature*) with direct implications on matters of human affairs, ecology, and economics.

Ecology is not just biological, but a human science as well. An early and influential social scientist in the history of human ecology was Herbert Spencer. Spencer was influenced by and reciprocated his influence onto the works of Charles Darwin. Herbert Spencer coined the phrase "survival of the fittest", he was an early founder of sociology where he developed the idea of society as an organism, and he created an early precedent for the socio-ecological approach that was the subsequent aim and link between sociology and human ecology.

Human ecology is the discipline that inquires into the patterns and process of interaction of humans with their environments. Human values, wealth, life-styles, resource use, and waste, etc. must affect and be affected by the physical and biotic environments along urban-rural gradients. The nature of these interactions is a legitimate ecological research topic and one of increasing importance.

The history of human ecology has strong roots in geography and sociology departments of the late 19th century. In this context a major historical development or landmark that stimulated research into the ecological relations between humans and their urban environments was founded in George Perkins Marsh's book *Man and Nature; or, physical geography as modified by human action*, which was published in 1864. Marsh was interested in the active agency of human-nature interactions (an early precursor to urban ecology or human niche construction) in frequent reference to the *economy of nature*.

In 1894, an influential sociologist at the University of Chicago named Albion W. Small, collaborated with sociologist George E. Vincent and published a ""laboratory guide" to studying people in their "every-day occupations."" This was a guidebook that trained students of sociology how they could study society in a way that a natural historian would study birds. Their publication "explicitly included the relation of the social world to the material environment."

The first English-language use of the term "ecology" is credited to American chemist and founder of the field of home economics, Ellen Swallow Richards. Richards first introduced the term as "oekology" in 1892, and subsequently developed the term "human ecology".

The term "human ecology" was published in 1907 in Ellen Swallow Richards work "Sanitation in Daily Life", defined there as "the study of the surroundings of human beings in the effects they produce on the lives of men". Richard's use of the term recognized humans as part of rather than separate from nature. The term made its first formal appearance in the field of sociology in the 1921 book "Introduction to the Science of Sociology", published by Robert E. Park and Ernest W. Burgess (also from the sociology department at the University of Chicago). Their student, Roderick D. McKenzie helped solidify human ecology as a sub-discipline within the Chicago school. These authors emphasized the difference between human ecology and ecology in general by highlighting cultural evolution in human societies.

Human ecology has a fragmented academic history with developments spread throughout a range of disciplines, including: home economics, geography, anthropology, sociology, zoology, and psychology. Some authors have argued that geography is human ecology. Much historical debate has hinged on the placement of humanity as part or as separate from nature. In light of the branching debate of what constitutes human ecology, recent interdisciplinary researchers have sought a unifying scientific field they have titled coupled human and natural systems that "builds on but moves beyond previous work (e.g., human ecology, ecological anthropology, environmental geography)." Other fields or branches related to the historical development of human ecology as a discipline include cultural ecology, urban ecology, environmental sociology, and anthropological ecology.

Biological ecologists have traditionally been reluctant to study human ecology gravitating instead to the allure of wild nature. Human ecology has a history of focusing attention on humans' impact on the biotic world. Paul Sears was an early proponent of applying human ecology, addressing topics aimed at the population explosion of humanity, global resource limits, pollution, and published a comprehensive account on human ecology as a discipline in 1954. He saw the vast "explosion" of problems humans were creating for the environment and reminded us that "what is important is the work to be done rather than the label." "When we as a profession learn to diagnose the total landscape, not only as the basis of our culture, but as an expression of it, and to share our special knowledge as widely as we can, we need not fear that our work will be ignored or that our efforts will be unappreciated."

Overview

Human ecology has been defined as a type of analysis applied to the relations in human beings that was traditionally applied to plants and animals in ecology. Toward this aim, human ecologists (which can include sociologists) integrate diverse perspectives from a broad spectrum of disciplines covering "wider points of view". In its 1972 premier edition, the editors of *Human Ecology: An Interdisciplinary Journal* gave an introductory statement on the scope of topics in human ecology. Their statement provides a broad overview on the interdisciplinary nature of the topic:

- Genetic, physiological, and social adaptation to the environment and to environmental change;
- The role of social, cultural, and psychological factors in the maintenance or disruption of ecosystems;
- Effects of population density on health, social organization, or environmental quality;

- New adaptive problems in urban environments;
- Interrelations of technological and environmental changes;
- The development of unifying principles in the study of biological and cultural adaptation;
- The genesis of maladaptions in human biological and cultural evolution;
- The relation of food quality and quantity to physical and intellectual performance and to demographic change;
- The application of computers, remote sensing devices, and other new tools and techniques

Forty years later in the same journal, Daniel G. Bates (2012) notes lines of continuity in the discipline and the way it has changed:

Today there is greater emphasis on the problems facing individuals and how actors deal with them with the consequence that there is much more attention to decision-making at the individual level as people strategize and optimize risk, costs and benefits within specific contexts. Rather than attempting to formulate a cultural ecology or even a specifically "human ecology" model, researchers more often draw on demographic, economic and evolutionary theory as well as upon models derived from field ecology.

While theoretical discussions continue, research published in *Human Ecology Review* suggests that recent discourse has shifted toward applying principles of human ecology. Some of these applications focus instead on addressing problems that cross disciplinary boundaries or transcend those boundaries altogether. Scholarship has increasingly tended away from Gerald L. Young's idea of a "unified theory" of human ecological knowledge—that human ecology may emerge as its own discipline—and more toward the pluralism best espoused by Paul Shepard: that human ecology is healthiest when "running out in all directions.". But human ecology is neither anti-discipline nor anti-theory, rather it is the ongoing attempt to formulate, synthesize, and apply theory to bridge the widening schism between man and nature. This new human ecology emphasizes complexity over reductionism, focuses on changes over stable states, and expands ecological concepts beyond plants and animals to include people.

Application to Epidemiology and Public Health

The application of ecological concepts to epidemiology has similar roots to those of other disciplinary applications, with Carl Linnaeus having played a seminal role. However, the term appears to have come into common use in the medical and public health literature in the mid-twentieth century. This was strengthened in 1971 by the publication of Epidemiology as Medical Ecology, and again in 1987 by the publication of a textbook on Public Health and Human Ecology. An "ecosystem health" perspective has emerged as a thematic movement, integrating research and practice from such fields as environmental management, public health, biodiversity, and economic development. Drawing in turn from the application of concepts such as the social-ecological model of health, human ecology has converged with the mainstream of global public health literature.

Connection to Home Economics

In addition to its links to other disciplines, human ecology has a strong historical linkage to the field of home economics through the work of Ellen Swallow Richards, among others. However, as early as the 1960s, a number of universities began to rename "home economics" departments, schools, and colleges as "human ecology" programs. In part, this name change was a response to perceived difficulties with the term "home economics" in a modernizing society, and reflects a recognition of "human ecology" as one of the initial choices for the discipline which was to become "home economics". Current Human Ecology programs include Cornell University College of Human Ecology and the University of Alberta's Department of Human Ecology, among others.

Niche of the Anthropocene

Perhaps the most important implication involves our view of human society. *Homo sapiens* is not an external disturbance, it is a keystone species within the system. In the long term, it may not be the magnitude of extracted goods and services that will determine sustainability. It may well be our disruption of ecological recovery and stability mechanisms that determines system collapse.

Changes to the Earth by human activities have been so great that a new geological epoch named the Anthropocene has been proposed. The human niche or ecological *polis* of human society, as it was known historically, has created entirely new arrangements of ecosystems as we convert matter into technology. Human ecology has created anthropogenic biomes (called anthromes). The habitats within these anthromes reach out through our road networks to create what has been called technoecosystems containing technosols. Technodiversity exists within these technoecosystems. In direct parallel to the concept of the ecosphere, human civilization has also created a technosphere. The way that the human species engineers or constructs technodiversity into the environment, threads back into the processes of cultural and biological evolution, including the human economy.

Ecosystem Services

The ecosystems of planet Earth are coupled to human environments. Ecosystems regulate the global geophysical cycles of energy, climate, soil nutrients, and water that in turn support and grow natural capital (including the environmental, physiological, cognitive, cultural, and spiritual dimensions of life). Ultimately, every manufactured product in human environments comes from natural systems. Ecosystems are considered common-pool resources because ecosystems do not exclude beneficiaries and they can be depleted or degraded. For example, green space within communities provides sustainable health services that reduces mortality and regulates the spread of vector borne disease. Research shows that people who are more engaged with regular access to natural areas have lower rates of diabetes, heart disease and psychological disorders. These ecological health services are regularly depleted through urban development projects that do not factor in the common-pool value of ecosystems.

A bumblebee pollinating a flower, one example of an ecosystem service

The ecological commons delivers a diverse supply of community services that sustains the well-being of human society. The Millennium Ecosystem Assessment, an international UN initiative involving more than 1,360 experts worldwide, identifies four main ecosystem service types having 30 sub-categories stemming from natural capital. The ecological commons includes provisioning (e.g., food, raw materials, medicine, water supplies), regulating (e.g., climate, water, soil retention, flood retention), cultural (e.g., science and education, artistic, spiritual), and supporting (e.g., soil formation, nutrient cycling, water cycling) services.

Policy and human institutions should rarely assume that human enterprise is benign. A safer assumption holds that human enterprise almost always exacts an ecological toll - a debit taken from the ecological commons.

Sixth Mass Extinction

Global assessments of biodiversity indicate that the current epoch, the Holocene (or Anthropocene) is a sixth mass extinction. Species loss is accelerating at 100–1000 times faster than average background rates in the fossil record. The field of conservation biology involves ecologists that are researching, confronting, and searching for solutions to sustain the planet's ecosystems for future generations.

"Human activities are associated directly or indirectly with nearly every aspect of the current extinction spasm."

Nature is a resilient system. Ecosystems regenerate, withstand, and are forever adapting to fluctuating environments. Ecological resilience is an important conceptual framework in conservation management and it is defined as the preservation of biological relations in ecosystems that persevere and regenerate in response to disturbance over time. However, persistent, systematic, large and nonrandom disturbance caused by the niche constructing behavior of human beings, habitat conversion and land development, has pushed many of the Earth's ecosystems to the extent of their resilient thresholds. Three planetary thresholds have already been crossed, including 1) biodiversity loss, 2) climate change, and 3) nitrogen cycles. These biophysical systems are ecologically interrelated and naturally resilient, but human civilization has transitioned the planet to an Anthropocene epoch, where the threshold for planetary scale resilience has been crossed and the ecological state of the Earth is deteriorating rapidly to the detriment of humanity. The world's fisheries and oceans, for example, are facing dire challenges as the threat of global collapse appears imminent, with serious ramifications for the well-being of humanity; while the Anthropocene is yet to be classified as an official epoch, current evidence suggest that "an epoch-scale boundary has been crossed within the last two centuries." The ecology of the planet is further threatened by global warming, but investments in nature conservation can provide a regulatory feedback to store and regulate carbon and other greenhouse gases.

Ecological Footprint

While we are used to thinking of cities as geographically discrete places, most of the land "occupied" by their residents lies far beyond their borders. The total area of land required to sustain an urban region (its "ecological footprint") is typically at least an order of magnitude greater than that contained within municipal boundaries or the associated built-up area.

In 1992, William Rees developed the ecological footprint concept. The ecological footprint and its close analog the water footprint has become a popular way of accounting for the level of impact that human society is imparting on the Earth's ecosystems. All indications are that the human enterprise is unsustainable as the footprint of society is placing too much stress on the ecology of the planet. The WWF 2008 living planet report and other researchers report that human civilization has exceeded the bio-regenerative capacity of the planet. This means that the footprint of human consumption is extracting more natural resources than can be replenished by ecosystems around the world.

Ecological Economics

Ecological economics is an economic science that extends its methods of valuation onto nature in an effort to address the inequity between market growth and biodiversity loss. Natural capital is the stock of materials or information stored in biodiversity that generates services that can enhance the welfare of communities. Population losses are the more sensitive indicator of natural capital than are species extinction in the accounting of ecosystem services. The prospect for recovery in the economic crisis of nature is grim. Populations, such as local ponds and patches of forest are being cleared away and lost at rates that exceed species extinctions. The mainstream growth-based economic system adopted by governments worldwide does not include a price or markets for natural capital. This type of economic system places further ecological debt onto future generations.

Many human-nature interactions occur indirectly due to the production and use of human-made (manufactured and synthesized) products, such as electronic appliances, furniture, plastics, airplanes, and automobiles. These products insulate humans from the natural environment, leading them to perceive less dependence on natural systems than is the case, but all manufactured products ultimately come from natural systems.

Human societies are increasingly being placed under stress as the ecological commons is diminished through an accounting system that has incorrectly assumed "... that nature is a fixed, indestructible capital asset." The current wave of threats, including massive extinction rates and concurrent loss of natural capital to the detriment of human society, is happening rapidly. This is called a biodiversity crisis, because 50% of the worlds species are predicted to go extinct within the next 50 years. Conventional monetary analyses are unable to detect or deal with these sorts of ecological problems. Multiple global ecological economic initiatives are being promoted to solve this problem. For example, governments of the G8 met in 2007 and set forth The Economics of Ecosystems and Biodiversity (TEEB) initiative:

In a global study we will initiate the process of analyzing the global economic benefit of biological diversity, the costs of the loss of biodiversity and the failure to take protective measures versus the costs of effective conservation.

The work of Kenneth E. Boulding is notable for building on the integration between ecology and its economic origins. Boulding drew parallels between ecology and economics, most generally in that they are both studies of individuals as members of a system, and indicated that the "household of man" and the "household of nature" could somehow be integrated to create a perspective of greater value.

Interdisciplinary Approaches

Human ecology may be defined: (1) from a bio-ecological standpoint as the study of man as the ecological dominant in plant and animal communities and systems; (2) from a bio-ecological standpoint as simply another animal affecting and being affected by his physical environment; and (3) as a human being, somehow different from animal life in general, interacting with physical and modified environments in a distinctive and creative way. A truly interdisciplinary human ecology will most likely address itself to all three.

Human ecology expands functionalism from ecology to the human mind. People's perception of a complex world is a function of their ability to be able to comprehend beyond the immediate, both in time and in space. This concept manifested in the popular slogan promoting sustainability: "think global, act local." Moreover, people's conception of community stems from not only their physical location but their mental and emotional connections and varies from "community as place, community as way of life, or community of collective action."

In these early years, human ecology was still deeply enmeshed in its respective disciplines: geography, sociology, anthropology, psychology, and economics. Scholars through the 1970s until present have called for a greater integration between all of the scattered disciplines that has each established formal ecological research.

In Art

While some of the early writers considered how art fit into a human ecology, it was Sears who posed the idea that in the long run human ecology will in fact look more like art. Bill Carpenter (1986) calls human ecology the "possibility of an aesthetic science," renewing dialogue about how art fits into a human ecological perspective. According to Carpenter, human ecology as an aesthetic science counters the disciplinary fragmentation of knowledge by examining human consciousness.

In Education

While the reputation of human ecology in institutions of higher learning is growing, there is no human ecology at the primary and secondary education levels. Educational theorist Sir Kenneth Robinson has called for diversification of education to promote creativity in academic and non-academic (i.e.- educate their "whole being") activities to implement a "new conception of human ecology".

Bioregionalism and Urban Ecology

In the late 1960s, ecological concepts started to become integrated into the applied fields, namely architecture, landscape architecture, and planning. Ian McHarg called for a future when all planning would be "human ecological planning" by default, always bound up in humans' relationships with their environments. He emphasized local, place-based planning that takes into consideration all the "layers" of information from geology to botany to zoology to cultural history. Proponents of the new urbanism movement, like James Howard Kunstler and Andres Duany, have embraced the term human ecology as way to describe the problem of—and prescribe the solutions for—the landscapes and lifestyles of an automobile oriented society. Duany has called the human ecology movement to be "the agenda for the years ahead." While McHargian planning is still widely respected, the landscape urbanism movement seeks a new understanding between human and environment relations. Among these theorists is Frederich Steiner, who published *Human Ecology: Following Nature's Lead* in 2002 which focuses on the relationships among landscape, culture, and planning. The work highlights the beauty of scientific inquiry by revealing those purely human dimensions which underlie our concepts of ecology. While Steiner discusses specific ecological settings, such as cityscapes and waterscapes, and the relationships between socio-cultural and environmental regions, he also takes a diverse approach to ecology—considering even the unique synthesis between ecology and political geography. Deiter Steiner's 2003 *Human Ecology: Fragments of Anti-fragmentary* view of the world is an important expose of recent trends in human ecology. Part literature review, the book is divided into four sections: "human ecology", "the implicit and the explicit", "structuration", and "the regional dimension". Much of the work stresses the need for transciplinarity, antidualism, and wholeness of perspective.

Biome

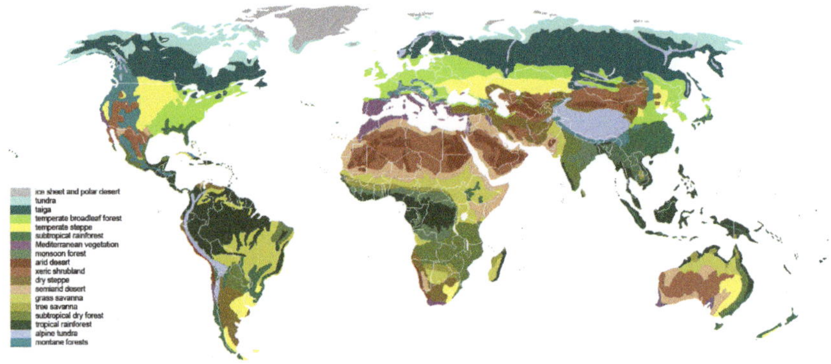

A map of different terrestrial biomes around the world.

A biome is a formation of plants and animals that have common characteristics due to similar climates and can be found over a range of continents. Spanning continents, Biomes are distinct biological communities that have formed in response to a shared physical climate. Biomes are distinct from habitats, because any biome can comprise a variety of habitats.

A biome contrasts with a microbiome. A microbiome is also a mix of organisms that coexist in a defined space, but on a much smaller scale. For example, the human microbiome is the collection of bacteria, viruses, and other microorganisms that are present on a human.

History of the Concept

The term was suggested in 1916 by Clements, originally as a synonym for biotic community of Möbius (1877). Later, it gained its current definition, based on earlier concepts of phytophysiognomy, formation and vegetation (used in opposition to flora), with the inclusion of the animal element and the exclusion of the taxonomic element of species composition. In 1935, Tansley would add the climatic and soil aspects to the idea, calling it ecosystem. The International Biological Program (1964-74) projects would popularize the concept of biome.

However, in some contexts, the term biome is used in a different manner. In German literature, particularly in the Walter terminology, the term is used similarly as biotope (a concrete geographycal unit), while the biome definition used in this article is used as a international, non-regional, terminology - irrespectively of the continent in which an area is present, it takes the same biome name - and corresponds to his "zonobiome", "orobiome" and "pedobiome" (biomes determinated by climate zone, altitude or soil).

In Brazilian literature, the term "biome" is sometimes used as synonym of "biogeographic province", an area based on species composition (the term "floristic province" being used when plant species are considered), or also as synonym of the "morphoclimatic and phytogeographical domain" of Ab'Sáber, a geographic space with subcontinental dimensions, with the predominance of similar geomorphologic and climatic characteristics, and of a certain vegetation form. Both includes many biomes in fact.

Classifications

Biomes are defined by climate regimes and biogeography.

A 1978 study on North American grasslands found a positive logistic correlation between evapotranspiration in mm/yr and above-ground net primary production in g/m^2/yr. The general results from the study were that precipitation and water use led to above-ground primary production, while solar irradiation and temperature lead to below-ground primary production (roots), and temperature and water lead to cool and warm season growth habit. These findings help explain the categories used in Holdridge's bioclassification scheme, which were then later simplified by Whittaker. The number of classification schemes and the variety of determinants used in those schemes, however, should be taken as strong indicators that biomes do not fit perfectly into the classification schemes created.

Holdridge (1947, 1964) Life Zones

Holdridge classified climates based on the biological effects of temperature and rainfall on vegetation under the assumption that these two abiotic factors are the largest determinants of the types of vegetation found in a habitat. Holdridge uses the four axes to define 30 so-called "humidity provinces", which are clearly visible in his diagram. While this scheme largely ignores soil and sun exposure, Holdridge acknowledged that these were important.

Allee (1949) Biome-types

The principal biome-types by Allee (1949):

- Tundra
- Taiga
- Deciduous forest
- Grasslands
- Desert
- High plateaus
- Tropical forest
- Minor terrestrial biomes

Kendeigh (1961) Biomes

The principal biomes of the world by Kendeigh (1961):

- Terrestrial
    - Temperate deciduous forest
    - Coniferous forest
    - Woodland
    - Chaparral
    - Tundra
    - Grassland
    - Desert
    - Tropical savanna
    - Tropical forest
- Marine
    - Oceanic plankton and nekton
    - Balanoid-gastropod-thallophyte

- o Pelecypod-annelid
- o Coral reef

Whittaker (1962, 1970, 1975) Biome-types

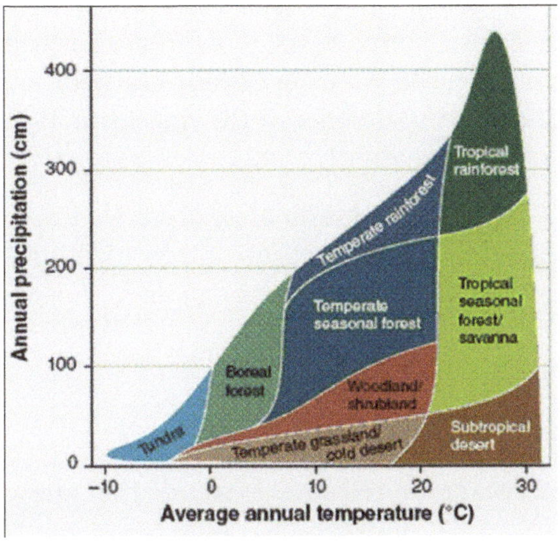

The distribution of vegetation types as a function of mean annual temperature and precipitation.

Whittaker classified biomes using two abiotic factors: precipitation and temperature. His scheme can be seen as a simplification of Holdridge's; more readily accessible, but missing Holdridge's greater specificity.

Whittaker based his approach on theoretical assertions and empirical sampling. He was in a unique position to make such a holistic assertion because he had previously compiled a review of biome classifications.

Key Definitions for Understanding Whittaker's Scheme

- Physiognomy: the apparent characteristics, outward features, or appearance of ecological communities or species.
- Biome: a grouping of terrestrial ecosystems on a given continent that are similar in vegetation structure, physiognomy, features of the environment and characteristics of their animal communities.
- Formation: a major kind of community of plants on a given continent.
- Biome-type: grouping of convergent biomes or formations of different continents, defined by physiognomy.
- Formation-type:a grouping of convergent formations.

Whittaker's distinction between biome and formation can be simplified: formation is used when applied to plant communities only, while biome is used when concerned with both plants and animals. Whittaker's convention of biome-type or formation-type is simply a broader method to categorize similar communities.

Whittaker's Parameters for Classifying Biome-types

Whittaker, seeing the need for a simpler way to express the relationship of community structure to the environment, used what he called "gradient analysis" of ecocline patterns to relate communities to climate on a worldwide scale. Whittaker considered four main ecoclines in the terrestrial realm.

1. Intertidal levels: The wetness gradient of areas that are exposed to alternating water and dryness with intensities that vary by location from high to low tide
2. Climatic moisture gradient
3. Temperature gradient by altitude
4. Temperature gradient by latitude

Along these gradients, Whittaker noted several trends that allowed him to qualitatively establish biome-types:

- The gradient runs from favorable to extreme, with corresponding changes in productivity.
- Changes in physiognomic complexity vary with how favorable of an environment exists (decreasing community structure and reduction of stratal differentiation as the environment becomes less favorable).
- Trends in diversity of structure follow trends in species diversity; alpha and beta species diversities decrease from favorable to extreme environments.
- Each growth-form (i.e. grasses, shrubs, etc.) has its characteristic place of maximum importance along the ecoclines.
- The same growth forms may be dominant in similar environments in widely different parts of the world.

Whittaker summed the effects of gradients (3) and (4) to get an overall temperature gradient, and combined this with gradient (2), the moisture gradient, to express the above conclusions in what is known as the Whittaker classification scheme. The scheme graphs average annual precipitation (x-axis) versus average annual temperature (y-axis) to classify biome-types.

Biome-types

- 1. Tropical rainforest

- 2. Tropical seasonal rainforest
    - deciduous
    - semideciduous
- 3. Temperate giant rainforest
- 4. Montane rainforest
- 5. Temperate deciduous forest
- 6. Temperate evergreen forest
    - needleleaf
    - sclerophyll
- 7. Subarctic-subalpin needle-leaved forests (taiga)
- 8. Elfin woodland
- 9. Thorn forests and woodlands
- 10. Thorn scrub
- 11. Temperate woodland
- 12. Temperate shrublands
    - deciduous
    - heath
    - sclerophyll
    - subalpine-needleleaf
    - subalpine-broadleaf
- 13. Savanna
- 14. Temperate grassland
- 15. Alpine grasslands
- 16. Tundra
- 17. Tropical desert
- 18. Warm-temperate desert

- 19. Cool temperate desert scrub
- 20. Arctic-alpine desert
- 21. Bog
- 22. Tropical fresh-water swamp forest
- 23. Temperate fresh-water swamp forest
- 24. Mangroove swamp
- 25. Saltmarsh

Goodall (1974-) Ecosystem Types

The multiauthored series *Ecosystems of the world*, edited by David W. Goodall, provides a comprehensive coverage of the major "ecosystem types or biomes" on earth:

- I. Terrestrial Ecosystems
    - A. Natural Terrestrial Ecosystems
        - 1. Wet Coastal Ecosystems
        - 2. Dry Coastal Ecosystems
        - 3. Polar and Alpine Tundra
        - 4. Mires: Swamp, Bog, Fen and Moor
        - 5. Temperate Deserts and Semi-Deserts
        - 6. Coniferous Forests
        - 7. Temperate Deciduous Forests
        - 8. Natural Grasslands
        - 9. Heathlands and Related Shrublands
        - 10. Temperate Broad-Leaved Evergreen Forests
        - 11. Mediterranean-Type Shrublands
        - 12. Hot Deserts and Arid Shrublands
        - 13. Tropical Savannas
        - 14. Tropical Rain Forest Ecosystems

- ☐ 15. Wetland Forests
- ☐ 16. Ecosystems of Disturbed Ground

o B. Managed Terrestrial Ecosystems

- ☐ 17. Managed Grasslands
- ☐ 18. Field Crop Ecosystems
- ☐ 19. Tree Crop Ecosystems
- ☐ 20. Greenhouse Ecosystems
- ☐ 21. Bioindustrial Ecosystems

• II. Aquatic Ecosystems

o A. Inland Aquatic Ecosystems

- ☐ 22. River and Stream Ecosystems
- ☐ 23. Lakes and Reservoirs

o B. Marine Ecosystems

- ☐ 24. Intertidal and Littoral Ecosystems
- ☐ 25. Coral Reefs
- ☐ 26. Estuaries and Enclosed Seas
- ☐ 27. Ecosystems of the Continental Shelves
- ☐ 28. Ecosystems of the Deep Ocean

o C. Managed Aquatic Ecosystems

- ☐ 29. Managed Aquatic Ecosystems

• III. Underground Ecosystems

o 30. Cave Ecosystems

Walter (1976, 2002) Zonobiomes

The eponymously-named Heinrich Walter classification scheme considers the seasonality of temperature and precipitation. The system, also assessing precipitation and temperature, finds nine major biome types, with the important climate traits and vegetation types. The boundaries of each biome correlate to the conditions of moisture and cold stress that are strong determinants of plant form, and therefore the vegetation that defines the region. Extreme conditions, such as flooding in a swamp, can create different kinds of communities within the same biome.

Zonobiome	Zonal soil type	Zonal vegetation type
ZB I. Equatorial, always moist, little temperature seasonality	Equatorial brown clays	Evergreen tropical rainforest
ZB II. Tropical, summer rainy season and cooler "winter" dry season	Red clays or red earths	Tropical seasonal forest, seasonal dry forest, scrub, or savanna
ZB III. Subtropical, highly seasonal, arid climate	Serosemes, sicrozemes	Desert vegetation with considerable exposed surface
ZB IV. Mediterranean, winter rainy season and summer drought	Mediterranean brown earths	Sclerophyllous (drought-adapted), frost-sensitive shrublands and woodlands
ZB V. Warm temperate, occasional frost, often with summer rainfall maximum	Yellow or red forest soils, slightly podsolic soils	Temperate evergreen forest, somewhat frost-sensitive
ZB VI. Nemoral, moderate climate with winter freezing	Forest brown earths and grey forest soils	Frost-resistant, deciduous, temperate forest
ZB VII. Continental, arid, with warm or hot summers and cold winters	Chernozems to serozems	Grasslands and temperate deserts
ZB VIII. Boreal, cold temperate with cool summers and long winters	Podsols	Evergreen, frost-hardy, needle-leaved forest (taiga)
ZB IX. Polar, short, cool summers and long, cold winters	Tundra humus soils with solifluction (permafrost soils)	Low, evergreen vegetation, without trees, growing over permanently frozen soils

Schültz (1988) Ecozones

Schültz (1988) defined nine ecozones.

- 1. polar/subpolar zone
- 2. boreal zone
- 3. humid mid-latitudes
- 4. arid mid-latitudes
- 5. tropical/subtropical arid lands

- 6. Mediterranean-type subtropics
- 7. seasonal tropics
- 8. humid subtropics
- 9. humid tropics

Bailey (1989) Ecoregions

Robert G. Bailey nearly developed a biogeographical classification system of ecoregions for the United States in a map published in 1976. He subsequently expanded the system to include the rest of North America in 1981, and the world in 1989. The Bailey system, based on climate, is divided into seven domains (polar, humid temperate, dry, humid, and humid tropical), with further divisions based on other climate characteristics (subarctic, warm temperate, hot temperate, and subtropical; marine and continental; lowland and mountain).

- 100 Polar Domain
    - 120 Tundra Division (Köppen: Ft)
    - M120 Tundra Division – Mountain Provinces
    - 130 Subarctic Division (Köppen: E)
    - M130 Subarctic Division – Mountain Provinces
- 200 Humid Temperate Domain
    - 210 Warm Continental Division (Köppen: portion of Dcb)
    - M210 Warm Continental Division – Mountain Provinces
    - 220 Hot Continental Division (Köppen: portion of Dca)
    - M220 Hot Continental Division – Mountain Provinces
    - 230 Subtropical Division (Köppen: portion of Cf)
    - M230 Subtropical Division – Mountain Provinces
    - 240 Marine Division (Köppen: Do)
    - M240 Marine Division – Mountain Provinces
    - 250 Prairie Division (Köppen: arid portions of Cf, Dca, Dcb)
    - 260 Mediterranean Division (Köppen: Cs)
    - M260 Mediterranean Division – Mountain Provinces

- 300 Dry Domain
    - 310 Tropical/Subtropical Steppe Division
    - M310 Tropical/Subtropical Steppe Division – Mountain Provinces
    - 320 Tropical/Subtropical Desert Division
    - 330 Temperate Steppe Division
    - 340 Temperate Desert Division
- 400 Humid Tropical Domain
    - 410 Savanna Division
    - 420 Rainforest Division

Olson & Dinerstein (1998) Biomes for WWF / Global 200

A team of biologists convened by the World Wildlife Fund (WWF) developed a scheme that divided the world's land area into biogeographic realms (called "ecozones" in a BBC scheme), and these into ecoregions (Olson & Dinerstein, 1998, etc.). Each ecoregion is characterized by a main biome (also called major habitat type).

This classification is used to define the Global 200 list of ecoregions identified by the WWF as priorities for conservation.

For the terrestrial ecoregions, there is a specific EcoID, format XXnnNN (XX is the biogeographic realm, nn is the biome number, NN is the individual number).

Biogeographic realms (terrestrial and freshwater)

- NA: Nearctic
- PA: Palearctic
- AT: Afrotropic
- IM: Indomalaya
- AA: Australasia
- NT: Neotropic
- OC: Oceania
- AN: Antarctic

It should be noted, however, that the applicability of the realms scheme above - based on Udvardy (1975) - to most freshwater taxa is unresolved.

Biogeographic Realms (Marine)

- Arctic
- Temperate Northern Atlantic
- Temperate Northern Pacific
- Tropical Atlantic
- Western Indo-Pacific
- Central Indo-Pacific
- Eastern Indo-Pacific
- Tropical Eastern Pacific
- Temperate South America
- Temperate Southern Africa
- Temperate Australasia
- Southern Ocean

Biomes (Terrestrial)

- 01: Tropical and subtropical moist broadleaf forests (tropical and subtropical, humid)
- 02: Tropical and subtropical dry broadleaf forests (tropical and subtropical, semihumid)
- 03: Tropical and subtropical coniferous forests (tropical and subtropical, semi-humid)
- 04: Temperate broadleaf and mixed forests (temperate, humid)
- 05: Temperate coniferous forests (temperate, humid to semihumid)
- 06: Boreal forests/taiga (subarctic, humid)
- 07: Tropical and subtropical grasslands, savannas, and shrublands (tropical and subtropical, semiarid)
- 08: Temperate grasslands, savannas, and shrublands (temperate, semiarid)
- 09: Flooded grasslands and savannas (temperate to tropical, fresh or brackish water inundated)

- 10: Montane grasslands and shrublands (alpine or montane climate)
- 11: Tundra (Arctic)
- 12: Mediterranean forests, woodlands, and scrub or sclerophyll forests (temperate warm, semihumid to semiarid with winter rainfall)
- 13: Deserts and xeric shrublands (temperate to tropical, arid)
- 14: Mangrove (subtropical and tropical, salt water inundated)

Biomes (Freshwater)

According to the WWF, the following are classified as freshwater biomes:

- Large lakes
- Large river deltas
- Polar freshwaters
- Montane freshwaters
- Temperate coastal rivers
- Temperate floodplain rivers and wetlands
- Temperate upland rivers
- Tropical and subtropical coastal rivers
- Tropical and subtropical floodplain rivers and wetlands
- Tropical and subtropical upland rivers
- Xeric freshwaters and endorheic basins
- Oceanic islands
- Streams and rivers

Biomes (Marine)

Biomes of the coastal and continental shelf areas (neritic zone):

- Polar
- Temperate shelves and sea
- Temperate upwelling
- Tropical upwelling
- Tropical coral

Summary of the Scheme

- Biosphere
    - Biogeographic realms (terrestrial) (8)
        - ☐ Ecoregions (867), each characterized by a main biome type (14)
            - ☐ Ecosystems (biotopes)
- Biosphere
    - Biogeographic realms (freshwater) (8)
        - ☐ Ecoregions (426), each characterized by a main biome type (12)
            - ☐ Ecosystems (biotopes)
- Biosphere
    - Biogeographic realms (marine) (13)
        - ☐ (Marine provinces) (62)
            - ☐ Ecoregions (232), each characterized by a main biome type (5)
                - ☐ Ecosystems (biotopes)

Example:

- Biosphere
    - Biogeographic realm: Palearctic
        - ☐ Ecoregion: Dinaric Mountains mixed forests (PA0418); biome type: temperate broadleaf and mixed forests
            - ☐ Ecosystem: Orjen, vegetation belt between 1,100–1,450 m, Oromediterranean zone, nemoral zone (temperate zone)
                - ☐ Biotope: *Oreoherzogio-Abietetum illyricae* Fuk. (Plant list)
                    - ☐ Plant: Silver fir (*Abies alba*)

Other Biomes

Marine Biomes

Pruvot (1896) zones or "systems":

- Litoral zone

- Pelagic zone
- Abyssal zone

Longhurst (1998) biomes:

- Coastal
- Polar
- Trade wind
- Westerly

Other marine habitat types (not covered yet by the Global 200/WWF scheme):

- Open sea
- Deep sea
- Hydrothermal vents
- Cold seeps
- Benthic zone
- Pelagic zone (trades and westerlies)
- Abyssal
- Hadal (ocean trench)
- Littoral/Intertidal zone
- Kelp forest
- Pack ice

Anthropogenic Biomes

Humans have altered global patterns of biodiversity and ecosystem processes. As a result, vegetation forms predicted by conventional biome systems can no longer be observed across much of Earth's land surface as they have been replaced by crop and rangelands or cities. Anthropogenic biomes provide an alternative view of the terrestrial biosphere based on global patterns of sustained direct human interaction with ecosystems, including agriculture, human settlements, urbanization, forestry and other uses of land. Anthropogenic biomes offer a new way forward in ecology and conservation by recognizing the irreversible coupling of human and ecological systems at global scales and moving us toward an understanding of how best to live in and manage our biosphere and the anthropogenic biomes we live in.

Major anthropogenic biomes:

- Dense settlements
- Croplands
- Rangelands
- Forested
- Indoor

Microbial Biomes

Endolithic Biomes

The endolithic biome, consisting entirely of microscopic life in rock pores and cracks, kilometers beneath the surface, has only recently been discovered, and does not fit well into most classification schemes.

Dermal Biome

The dermal biome is the living ecosystem that animals (including humans) have evolved, that permits them to live symbiotically and in balance with the microbes on and in them (the microbiome). This ecosystem consists of skin, follicles, hair, sebaceous glands, sweat glands, arrector pili muscles, peptides, proteins, lipids and its associated microbiota. A healthy dermal biome has several functions: it resists infection of pathogens, protects against moisture loss and water damage, dynamically regulates body temperature and supports the healthy renewal of skin through the epidermal cell life cycle.

- Infection Resistance: Commensal microbiota assist the dermal biome resist infection for pathogenic bacteria by i] out-competing pathogens for resources, ii] training or stimulating the host's immune system to defeat the pathogen, or iii] expressing substances that are directly hostile to the pathogen.
- Water-barrier regulation: The dermal biome regulates the water barrier – preventing moisture from escaping (except when expressed as sweat) and preventing environmental water from permeating the skin. Because environmental water can have a chilling effect to mammals and warm-blooded animals when kept in close proximity to the epidermis, the dermal biome also produces hydrophobic lipids that repel water.
- Temperature regulation: The dermal biome is responsible for thermoregulation. To regulate excess heat, the dermal biome activates the sweat glands, allowing for evaporative cooling as sweat evaporates. To regulate cooler temperatures, the arrector pili muscles contract, causing hairs to "stand up" (goosebumps), and thereby trap an insulating blanket of air close to the skin.

- Skin renewal: a healthy biome supports the replacement of skin through the life cycle of epidermal cells as they proliferate in the basal layers of the epidermis until they die and are shed (desquamation).

Ecological Niche

The flightless dung beetle occupies a niche exploiting animal droppings as a food source

In ecology, a niche is a term with a variety of meanings related to the behavior of a species living under specific environmental conditions. The ecological niche describes how an organism or population responds to the distribution of resources and competitors (for example, by growing when resources are abundant, and when predators, parasites and pathogens are scarce) and how it in turn alters those same factors (for example, limiting access to resources by other organisms, acting as a food source for predators and a consumer of prey). "The type and number of variables comprising the dimensions of an environmental niche vary from one species to another [and] the relative importance of particular environmental variables for a species may vary according to the geographic and biotic contexts".

The notion of ecological niche is central to ecological biogeography, which focuses on spatial patterns of ecological communities. "Species distributions and their dynamics over time result from properties of the species, environmental variation..., and interactions between the two — in particular the abilities of some species, especially our own, to modify their environments and alter the range dynamics of many other species." Alteration of an ecological niche by its inhabitants is the topic of niche construction.

The majority of species exist in a standard ecological niche, sharing behaviors and adaptations similar to the other closely related species within the same broad taxonomic class, but there are exceptions. A premier example of a non-standard niche filling species is the flightless, ground-dwelling kiwi bird of New Zealand, which feeds on worms and other ground creatures, and lives its life in a mammal niche. Island biogeography can help explain island species and associated unfilled niches.

Grinnellian Niche

The ecological meaning of niche comes from the meaning of niche as a recess in a wall for a statue, which itself is probably derived from the Middle French word *nicher*, meaning *to nest*. The term was coined by the naturalist Roswell Hill Johnson but Joseph Grinnell was probably the first to use it in a research program in 1917, in his paper "The niche relationships of the California Thrasher".

The Grinnellian niche concept embodies the idea that the niche of a species is determined by the habitat in which it lives and its accompanying behavioral adaptations. In other words, the niche is the sum of the habitat requirements and behaviors that allow a species to persist and produce offspring. For example, the behavior of the California thrasher is consistent with the chaparral habitat it lives in—it breeds and feeds in the underbrush and escapes from its predators by shuffling from underbrush to underbrush. Its 'niche' is defined by the felicitous complementing of the thrasher's behavior and physical traits (camouflaging color, short wings, strong legs) with this habitat.

This perspective of niche allows for the existence of both ecological equivalents and empty niches. An ecological equivalent to an organism is an organism from a different taxonomic group exhibiting similar adaptations in a similar habitat, an example being the different succulents found in American and African deserts, cactus and euphorbia. As another example, the *Anolis* lizards of the Greater Antilles are a rare example of convergent evolution, adaptive radiation, and the existence of ecological equivalents: the *Anolis* lizards evolved in similar microhabitats independently of each other and resulted in the same ecomorphs across all four islands.

Eltonian Niche

In 1927 Charles Sutherland Elton, a British ecologist, defined a niche as follows: "The 'niche' of an animal means its place in the biotic environment, *its relations to food and enemies*."

Elton classified niches according to foraging activities ("food habits"): "For instance there is the niche that is filled by birds of prey which eat small animals such as shrews and mice. In an oak wood this niche is filled by tawny owls, while in the open grassland it is occupied by kestrels. The existence of this carnivore niche is dependent on the further fact that mice form a definite herbivore niche in many different associations, although the actual species of mice may be quite different."

Conceptually, the Eltonian niche introduces the idea of a species' *response to* and *effect on* the environment. Unlike other niche concepts the Eltonian niche emphasizes that a species not only grows in and responds to an environment based on available resources, predators, and climatic conditions, a species may also change the availability and behavior of those factors as it grows. In an extreme example, beavers require certain resources in order to survive and reproduce, but also construct dams that alter water flow in the river where the beaver lives. Thus, the beaver affects the biotic and abiotic conditions of other species that live in and near the watershed. In a more subtle case, competitors that consume resources at different rates can lead to cycles in resource density that differ between species. Not only do species grow differently with respect to resource density, their own population growth can lead to different effects on resource density over time.

Hutchinsonian Niche

The shape of the bill of this purple-throated carib is complementary to the shape of the flower, enabling it to exploit the nectar as a resource.

The Hutchinsonian niche is an n-dimensional hypervolume, where the dimensions are environmental conditions and resources, that define the requirements of an individual or a species to practice "its" way of life, more particularly, for its population to persist. The "hypervolume" defines the multi-dimensional space of resources (e.g., light, nutrients, structure, etc.) available to (and specifically used by) organisms, and "all species other than those under consideration are regarded as part of the coordinate system."

The niche concept was popularized by the zoologist G. Evelyn Hutchinson in 1957. Hutchinson inquired into the question of why there are so many types of organisms in any one habitat. His work inspired many others to develop models to explain how many and how similar coexisting species could be within a given community, and led to the concepts of 'niche breadth' (the variety of resources or habitats used by a given species), 'niche partitioning' (resource differentiation by coexisting species), and 'niche overlap' (overlap of resource use by different species).

Statistics were introduced into the Hutchinson niche by Robert MacArthur and Richard Levins using the 'resource-utilization' niche employing histograms to describe the 'frequency of occurrence' as a function of a Hutchinson coordinate. So, for instance, a Gaussian might describe the frequency with which a species ate prey of a certain size, giving a more detailed niche description than simply specifying some median or average prey size. For such a bell-shaped distribution, the *position*, *width* and *form* of the niche correspond to the *mean*, *standard deviation* and the actual distribution itself. One advantage in using statistics is illustrated in the figure, where it is clear that for the narrower distributions (top) there is no competition for prey between the extreme left and extreme right species, while for the broader distribution (bottom), niche overlap indicates competition can occur between all species. The resource-utilization approach consists in postulating that not only competition *can* occur, but also that it *does* occur, and that overlap in resource utilization directly enables the estimation of the competition coefficients. This postulate, however, can be misguided, as it ignores the impacts that the resources of each category have on the organism and the impacts that the organism has on the resources of each category. For instance, the resource in the overlap region can be non-limiting, in which case there is no competition for this resource despite niche overlap.

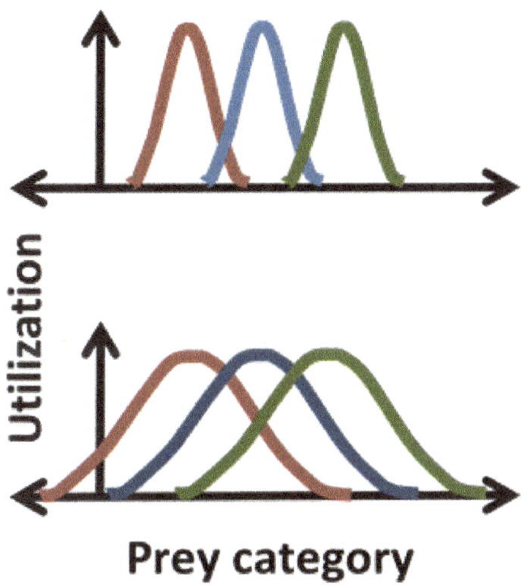

Where three species eat some of the same prey, a statistical picture of each niche shows overlap in resource usage between three species, indicating where competition is strongest.

An organism free of interference from other species could use the full range of conditions (biotic and abiotic) and resources in which it could survive and reproduce which is called its fundamental niche. However, as a result of pressure from, and interactions with, other organisms (i.e. inter-specific competition) species are usually forced to occupy a niche that is narrower than this, and to which they are mostly highly adapted; this is termed the realized niche. Hutchinson used the idea of competition for resources as the primary mechanism driving ecology, but overemphasis upon this focus has proved to be a handicap for the niche concept. In particular, overemphasis upon a species' dependence upon resources has led to too little emphasis upon the effects of organisms on their environment, for instance, colonization and invasions.

The term "adaptive zone" was coined by the paleontologist George Gaylord Simpson to explain how a population could jump from one niche to another that suited it, jump to an 'adaptive zone', made available by virtue of some modification, or possibly a change in the food chain, that made the adaptive zone available to it without a discontinuity in its way of life because the group was 'pre-adapted' to the new ecological opportunity.

As a hemi-parasitic plant, the mistletoe in this tree exploits its host for nutrients and as a place to grow.

Hutchinson's "niche" (a description of the ecological space occupied by a species) is subtly different from the "niche" as defined by Grinnell (an ecological role, that may or may not be actually filled by a species).

A niche is a very specific segment of ecospace occupied by a single species. On the presumption that no two species are identical in all respects (called Hardin's 'axiom of inequality') and the competitive exclusion principle, *some* resource or adaptive dimension will provide a niche specific to each species. Species can however share a 'mode of life' or 'autecological strategy' which are broader definitions of ecospace. For example, Australian grasslands species, though different from those of the Great Plains grasslands, exhibit similar modes of life.

Once a niche is left vacant, other organisms can fill that position. For example, the niche that was left vacant by the extinction of the tarpan has been filled by other animals (in particular a small horse breed, the konik). Also, when plants and animals are introduced into a new environment, they have the potential to occupy or invade the niche or niches of native organisms, often outcompeting the indigenous species. Introduction of non-indigenous species to non-native habitats by humans often results in biological pollution by the exotic or invasive species.

The mathematical representation of a species' fundamental niche in ecological space, and its subsequent projection back into geographic space, is the domain of niche modelling.

Parameters

The different dimensions, or plot axes, of a niche represent different biotic and abiotic variables. These factors may include descriptions of the organism's life history, habitat, trophic position (place in the food chain), and geographic range. According to the competitive exclusion principle, no two species can occupy the same niche in the same environment for a long time. The parameters of a realized niche are described by the realized niche width of that species.

Biomass (Ecology)

Apart from bacteria, the total global live biomass has been estimated as 560 billion tonnes C, most of which is found in forests.

Shallow aquatic environments, such as wetlands, estuaries and coral reefs, can be as productive as forests, generating similar amounts of new biomass each year on a given area.

Biomass, in ecology, is the mass of living biological organisms in a given area or ecosystem at a given time. Biomass can refer to *species biomass*, which is the mass of one or more species, or to *community biomass*, which is the mass of all species in the community. It can include microorganisms, plants or animals. The mass can be expressed as the average mass per unit area, or as the total mass in the community.

How biomass is measured depends on why it is being measured. Sometimes, the biomass is regarded as the natural mass of organisms *in situ*, just as they are. For example, in a salmon fishery, the salmon biomass might be regarded as the total wet weight the salmon would have if they were taken out of the water. In other contexts, biomass can be measured in terms of the dried organic mass, so perhaps only 30% of the actual weight might count, the rest being water. For other purposes, only biological tissues count, and teeth, bones and shells are excluded. In some applications, biomass is measured as the mass of organically bound carbon (C) that is present.

Apart from bacteria, the total live biomass on Earth is about 560 billion tonnes C, and the total annual primary production of biomass is just over 100 billion tonnes C/yr. The total live biomass of bacteria may be as much as that of plants and animals or may be much less. The total amount of DNA base pairs on Earth, as a possible approximation of global biodiversity, is estimated at 5.0×10^{37}, and weighs 50 billion tonnes. In comparison, the total mass of the biosphere has been estimated to be as much as 4 TtC (trillion tons of carbon).

Ecological Pyramids

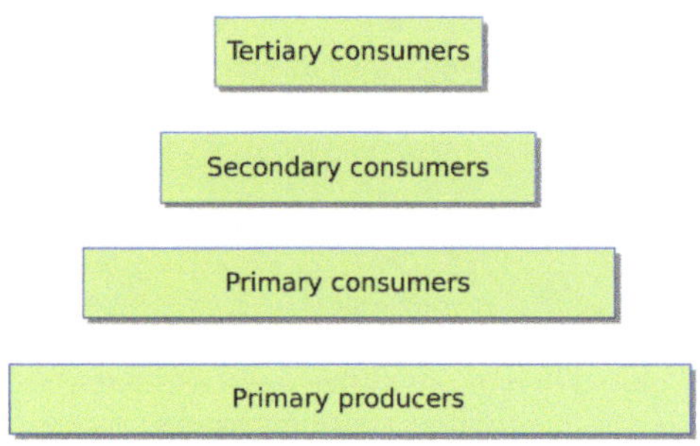

An ecological pyramid.

An ecological pyramid is a graphical representation that shows, for a given ecosystem, the relationship between biomass or biological productivity and trophic levels.

- A *biomass pyramid* shows the amount of biomass at each trophic level.
- A *productivity pyramid* shows the production or turn-over in biomass at each trophic level.

An ecological pyramid provides a snapshot in time of an ecological community.

The bottom of the pyramid represents the primary producers (autotrophs). The primary producers take energy from the environment in the form of sunlight or inorganic chemicals and use it to create energy-rich molecules such as carbohydrates. This mechanism is called primary production. The pyramid then proceeds through the various trophic levels to the apex predators at the top.

When energy is transferred from one trophic level to the next, typically only ten percent is used to build new biomass. The remaining ninety percent goes to metabolic processes or is dissipated as heat. This energy loss means that productivity pyramids are never inverted, and generally limits food chains to about six levels. However, in oceans, biomass pyramids can be wholly or partially inverted, with more biomass at higher levels.

Terrestrial Biomass

Terrestrial biomass generally decreases markedly at each higher trophic level (plants, herbivores, carnivores). Examples of terrestrial producers are grasses, trees and shrubs. These have a much higher biomass than the animals that consume them, such as deer, zebras and insects. The level with the least biomass are the highest predators in the food chain, such as foxes and eagles.

In a temperate grassland, grasses and other plants are the primary producers at the bottom of the pyramid. Then come the primary consumers, such as grasshoppers, voles and bison, followed by the secondary consumers, shrews, hawks and small cats. Finally the tertiary consumers, large cats and wolves. The biomass pyramid decreases markedly at each higher level.

Ocean Biomass

The Marine Food Chain

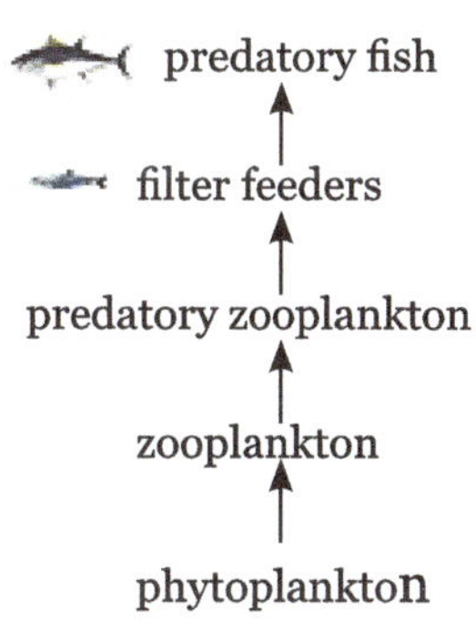

Ocean biomass, in a reversal of terrestrial biomass, can increase at higher trophic levels. In the ocean, the food chain typically starts with phytoplankton, and follows the course:

Phytoplankton → zooplankton → predatory zooplankton → filter feeders → predatory fish

Phytoplankton are the main primary producers at the bottom of the marine food chain. Phytoplankton use photosynthesis to convert inorganic carbon into protoplasm. They are then consumed by microscopic animals called zooplankton.

Zooplankton comprise the second level in the food chain, and includes small crustaceans, such as copepods and krill, and the larva of fish, squid, lobsters and crabs.

In turn, small zooplankton are consumed by both larger predatory zooplankters, such as krill, and by forage fish, which are small schooling filter feeding fish. This makes up the third level in the food chain.

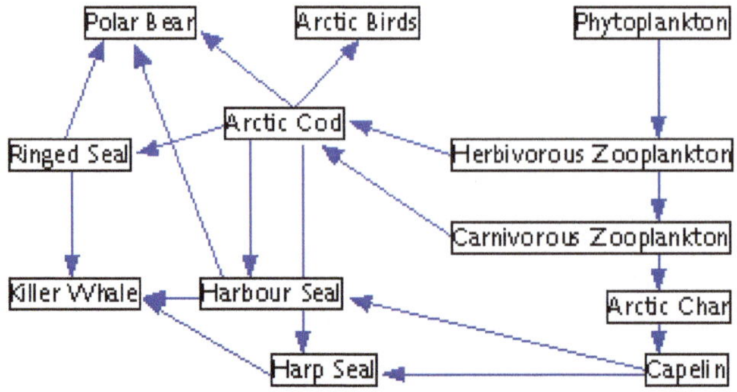

An ocean food web showing a network of food chains

The fourth trophic level consists of predatory fish, marine mammals and seabirds that consume forage fish. Examples are swordfish, seals and gannets.

Apex predators, such as orcas, which can consume seals, and shortfin mako sharks, which can consume swordfish, make up the fifth trophic level. Baleen whales can consume zooplankton and krill directly, leading to a food chain with only three or four trophic levels.

Marine environments can have inverted biomass pyramids. In particular, the biomass of consumers (copepods, krill, shrimp, forage fish) is larger than the biomass of primary producers. This happens because the ocean's primary producers are tiny phytoplankton that grow and reproduce rapidly, so a small mass can have a fast rate of primary production. In contrast, terrestrial primary producers grow and reproduce slowly.

There is an exception with cyanobacteria. Marine cyanobacteria are the smallest known photosynthetic organisms; the smallest of all, *Prochlorococcus*, is just 0.5 to 0.8 micrometres across. Prochlorococcus is possibly the most plentiful species on Earth: a single millilitre of surface seawater may contain 100,000 cells or more. Worldwide, there are estimated to be several octillion ($\sim 10^{27}$) individuals. *Prochlorococcus* is ubiquitous between 40°N and 40°S and dominates in the oligotrophic (nutrient poor) regions of the oceans. The bacterium accounts for an estimated 20% of the oxygen in the Earth's atmosphere, and forms part of the base of the ocean food chain.

Bacterial Biomass

There are typically 50 million bacterial cells in a gram of soil and a million bacterial cells in a millilitre of fresh water. In a much-cited study from 1998 the world bacterial biomass was calculated to be 350 to 550 billions of tonnes of carbon, equal to between 60% and 100% of the carbon in plants. More recent studies of seafloor microbes have cast considerable doubt on that, one study in 2012 reduced the calculated microbial biomass on the seafloor from the original 303 billions of tonnes of C to just 4.1 billions of tonnes of C, reducing the global biomass of prokaryotes to 50 to 250 billions of tonnes of C. Further, if the average per cell biomass of prokaryotes is reduced from 86 to 14 femtograms C then the global biomass of prokaryotes is reduced to 13 to 44.5 billions of tonnes of C, equal to between 2.4% and 8.1% of the carbon in plants.

Geographic location	Number of cells ($\times 10^{29}$)	Billions of tonnes of carbon
Ocean floor	2.9 to 50	4.1 to 303
Open ocean	1.2	1.7 to 10
Terrestrial soil	2.6	3.7 to 22
Subsurface terrestrial	2.5 to 25	3.5 to 215

Global Biomass

Estimates for the global biomass of species and higher level groups are not always consistent across the literature. Apart from bacteria, the total global biomass has been estimated at about 560 billion tonnes C. Most of this biomass is found on land, with only 5 to 10 billion tonnes C found in the oceans. On land, there is about 1,000 times more plant biomass (*phytomass*) than animal biomass (*zoomass*). About 18% of this plant biomass is eaten by the land animals. However, in the ocean, the animal biomass is nearly 30 times larger than the plant biomass. Most ocean plant biomass is eaten by the ocean animals.

	name	number of species	date of estimate	individual count	mean living mass of individual	percent biomass (dried)	total number of carbon atoms	global dry biomass in million tonnes	global wet (fresh) biomass in million tonnes
Terrestrial	Humans	1	2012	7.0 billion	50 kg (incl children)	30%	3.5×10^{26}	105	350
			2005	4.63 billion	62 kg (excl children)				287
	Cattle	1		1.3 billion	400 kg	30%		156	520
	Sheep and goats	2	2002	1.75 billion	60 kg	30%		31.5	105
	Chickens	1		24 billion	2 kg	30%		14.4	48
	Ants	12,649		10^{7} - 10^{8} billion	3×10^{-6} kg (0.003 grams)	30%		10–100	30-300
	Termites	>2,800	1996						445

Marine	Blue whales	1	Pre-whaling	340,000		40%			36
			2001	4700		40%			0.5
	Fish	>10,000	2009					800-2,000	
	Antarctic krill	1	1924–2004	7.8×10^{14}	0.486 g				379
	Copepods (a zooplankton)	13,000			10^{-6} - 10^{-9} kg		1×10^{37}		
	Cyanobacteria (a picoplankton)	?	2003						1,000
Global	Prokaryotes (bacteria)	?	1998	$4 - 6 \times 10^{30}$ cells			1.76-2.76×10^{40}	350,000-550,000	

Humans comprise about 100 million tonnes of the Earth's dry biomass, domesticated animals about 700 million tonnes, and crops about 2 billion tonnes. The most successful animal species, in terms of biomass, may well be Antarctic krill, *Euphausia superba*, with a fresh biomass approaching 500 million tonnes, although domestic cattle may also reach these immense figures. However, as a group, the small aquatic crustaceans called copepods may form the largest animal biomass on earth. A 2009 paper in *Science* estimates, for the first time, the total world fish biomass as somewhere between 0.8 and 2.0 billion tonnes. It has been estimated that about 1% of the global biomass is due to phytoplankton, and fully 25% is due to fungi.

Grasses, trees and shrubs have a much higher biomass than the animals that consume them

The total biomass of bacteria may equal that of plants.

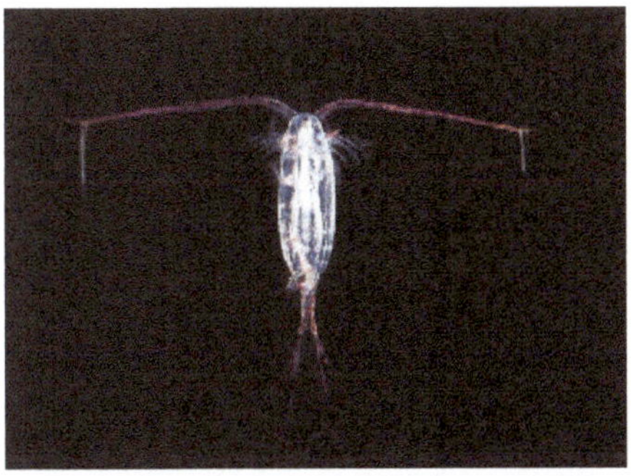

Copepods may form the largest biomass of any animal species group.

Antarctic krill form one of the largest biomasses of any individual animal species.

It has been claimed that fungi make up 25% of the global biomass

Global Rate of Production

Net primary production is the rate at which new biomass is generated, mainly due to photosynthesis. Global primary production can be estimated from satellite observations. Satellites scan the normalised difference vegetation index (NDVI) over terrestrial habitats, and scan sea-surface chlorophyll levels over oceans. This results in 56.4 billion tonnes C/yr (53.8%), for terrestrial primary production, and 48.5 billion tonnes C/yr for oceanic primary production. Thus, the total photoautotrophic primary production for the Earth is about 104.9 billion tonnes C/yr. This translates to about 426 $gC/m^2/yr$ for land production (excluding areas with permanent ice cover), and 140 $gC/m^2/yr$ for the oceans.

However, there is a much more significant difference in standing stocks—while accounting for almost half of total annual production, oceanic autotrophs account for only about 0.2% of the total biomass. Autotrophs may have the highest global proportion of biomass, but they are closely rivaled or surpassed by microbes.

Terrestrial freshwater ecosystems generate about 1.5% of the global net primary production.

Some global producers of biomass in order of productivity rates are

Producer	Biomass productivity ($gC/m^2/yr$)	Ref	Total area (million km^2)	Ref	Total production (billion tonnes C/yr)
Swamps and Marshes	2,500				
Tropical rainforests	2,000		8		16
Coral reefs	2,000		0.28		0.56
Algal beds	2,000				
River estuaries	1,800				
Temperate forests	1,250		19		24
Cultivated lands	650		17		11
Tundras	140				
Open ocean	125		311		39
Deserts	3		50		0.15

References

- Eric Laferrière; Peter J. Stoett (2 September 2003). International Relations Theory and Ecological Thought: Towards a Synthesis. Routledge. pp. 25–. ISBN 978-1-134-71068-3.
- O'Neill, D. L.; Deangelis, D. L.; Waide, J. B.; Allen, T. F. H. (1986). A Hierarchical Concept of Ecosystems. Princeton University Press. p. 253. ISBN 0-691-08436-X.
- Levin, S. A. (1999). Fragile Dominion: Complexity and the Commons. Reading, MA: Perseus Books. ISBN 978-0-7382-0319-5.

- Noss, R. F.; Carpenter, A. Y. (1994). Saving Nature's Legacy: Protecting and Restoring Biodiversity. Island Press. p. 443. ISBN 978-1-55963-248-5.
- Begon, M.; Townsend, C. R.; Harper, J. L. (2005). Ecology: From Individuals to Ecosystems (4th ed.). Wiley-Blackwell. p. 752. ISBN 1-4051-1117-8.
- Vandermeer, J. H.; Goldberg, D. E. (2003). Population Ecology: First Principles. Woodstock, Oxfordshire: Princeton University Press. ISBN 0-691-11440-4.
- Levins, R. (1970). "Extinction". In Gerstenhaber, M. Some Mathematical Questions in Biology. pp. 77–107. ISBN 978-0-8218-1152-8.
- Dingle, H. (1996-01-18). Migration: The Biology of Life on the Move. Oxford University Press. p. 480. ISBN 0-19-509723-8.
- Hanski, I.; Gaggiotti, O. E., eds. (2004). Ecology, Genetics and Evolution of Metapopulations. Burlington, MA: Elsevier Academic Press. ISBN 0-12-323448-4.
- MacKenzie; D.I. (2006). Occupancy Estimation and Modeling: Inferring Patterns and Dynamics of Species Occurrence. London, UK: Elsevier Academic Press. p. 324. ISBN 978-0-12-088766-8.
- Allee, W. C.; Park, O.; Emerson, A. E.; Park, T.; Schmidt, K. P. (1949). Principles of Animal Ecology. W. B. Sunders, Co. p. 837. ISBN 0-7216-1120-6.
- Wilson, E. O. (2000). Sociobiology: The New Synthesis (25th anniversary ed.). President and Fellows of Harvard College. ISBN 978-0-674-00089-6.
- Gilbert, F. S. (1990). Insect life cycles: Genetics, evolution, and co-ordination. New York, NY: Springer-Verlag. p. 258. ISBN 0-387-19550-5.
- Francisco J Varela; Evan Thompson; Eleanor Rosch (1993). The Embodied Mind: Cognitive Science and Human Experience (Paperback ed.). MIT Press. p. 174. ISBN 9780262261234.
- Campbell, Neil A.; Williamson, Brad; Heyden, Robin J. (2006). Biology: Exploring Life. Boston, Massachusetts: Pearson Prentice Hall. ISBN 0-13-250882-6.
- Cronk, J. K.; Fennessy, M. S. (2001). Wetland Plants: Biology and Ecology. Washington, D.C.: Lewis Publishers. ISBN 1-56670-372-7.
- Wills, C.; Bada, J. (2001). The Spark of Life: Darwin and the Primeval Soup. Cambridge, MA: Perseus Publishing. ISBN 978-0-7382-0493-2.
- Rory Putman; Stephen D. Wratten (1984). "§5.2 Parameters of the niche". Principles of ecology. University of California Press. p. 107. ISBN 9780520052543.

5

Protection of Biosphere

Biodiversity refers to all life on Earth. It is a measure of all organisms that exist in different ecosystems. The other aspects considered are environmental protection, habitat conservation, conservation biology and conservation movement. The aim of this chapter is to explore all the aspects of biosphere. These topics are crucial for a complete understanding of the subject.

Biodiversity

variability of life on Earth. One of the most widely used definitions defines it in terms of the variability within species, between species and between ecosystems. It is a measure of the variety of organisms present in different ecosystems. This can refer to genetic variation, ecosystem variation, or species variation (number of species) within an area, biome, or planet. Terrestrial biodiversity tends to be greater near the equator, which seems to be the result of the warm climate and high primary productivity. Biodiversity is not distributed evenly on Earth. It is richest in the tropics. Marine biodiversity tends to be highest along coasts in the Western Pacific, where sea surface temperature is highest and in the mid-latitudinal band in all oceans. There are latitudinal gradients in species diversity. Biodiversity generally tends to cluster in hotspots, and has been increasing through time, but will be likely to slow in the future.

The number and variety of plants, animals and other organisms that exist is known as biodiversity. It is an essential component of nature and it ensures the survival of human species by providing food, fuel, shelter, medicines and other resources to mankind. The richness of biodiversity depends on the climatic conditions and area of the region. All species of plants taken together are known as flora and about 70,000 species of plants are known to date. All species of animals taken together are known as fauna which includes birds, mammals, fish, reptiles, insects, crustaceans, molluscs, etc.

Rapid environmental changes typically cause mass extinctions. More than 99 percent of all species, amounting to over five billion species, that ever lived on Earth are estimated to be extinct. Estimates on the number of Earth's current species range from 10 million to 14 million, of which about 1.2 million have been documented and over 86 percent have not yet been described. More recently, in May 2016, scientists reported that 1 trillion species are estimated to be on Earth currently with only one-thousandth of one percent described. The total amount of related DNA base pairs on Earth is estimated at 5.0 x 10^{37} and weighs 50 billion tonnes. In comparison, the total mass of the biosphere has been estimated to be as much as 4 TtC (trillion tons of carbon). In July 2016, scientists reported identifying a set of 355 genes from the Last Universal Common Ancestor (LUCA) of all organisms living on Earth.

The age of the Earth is about 4.54 billion years old. The earliest undisputed evidence of life on Earth dates at least from 3.5 billion years ago, during the Eoarchean Era after a geological crust started to solidify following the earlier molten Hadean Eon. There are microbial mat fossils found in 3.48 billion-year-old sandstone discovered in Western Australia. Other early physical evidence of a biogenic substance is graphite in 3.7 billion-year-old meta-sedimentary rocks discovered in Western Greenland. More recently, in 2015, "remains of biotic life" were found in 4.1 billion-year-old rocks in Western Australia. According to one of the researchers, "If life arose relatively quickly on Earth .. then it could be common in the universe."

Since life began on Earth, five major mass extinctions and several minor events have led to large and sudden drops in biodiversity. The Phanerozoic eon (the last 540 million years) marked a rapid growth in biodiversity via the Cambrian explosion—a period during which the majority of multicellular phyla first appeared. The next 400 million years included repeated, massive biodiversity losses classified as mass extinction events. In the Carboniferous, rainforest collapse led to a great loss of plant and animal life. The Permian–Triassic extinction event, 251 million years ago, was the worst; vertebrate recovery took 30 million years. The most recent, the Cretaceous–Paleogene extinction event, occurred 65 million years ago and has often attracted more attention than others because it resulted in the extinction of the dinosaurs.

The period since the emergence of humans has displayed an ongoing biodiversity reduction and an accompanying loss of genetic diversity. Named the Holocene extinction, the reduction is caused primarily by human impacts, particularly habitat destruction. Conversely, biodiversity impacts human health in a number of ways, both positively and negatively.

The United Nations designated 2011–2020 as the United Nations Decade on Biodiversity.

Etymology

The term biological diversity was used first by wildlife scientist and conservationist Raymond F. Dasmann in the year 1968 lay book *A Different Kind of Country* advocating conservation. The term was widely adopted only after more than a decade, when in the 1980s it came into common usage in science and environmental policy. Thomas Lovejoy, in the foreword to the book *Conservation Biology*, introduced the term to the scientific community. Until then the term "natural diversity" was common, introduced by The Science Division of The Nature Conservancy in an important 1975 study, "The Preservation of Natural Diversity." By the early 1980s TNC's Science program and its head, Robert E. Jenkins, Lovejoy and other leading conservation scientists at the time in America advocated the use of the term "biological diversity".

The term's contracted form *biodiversity* may have been coined by W.G. Rosen in 1985 while planning the 1986 *National Forum on Biological Diversity* organized by the National Research Council (NRC). It first appeared in a publication in 1988 when sociobiologist E. O. Wilson used it as the title of the proceedings of that forum.

Since this period the term has achieved widespread use among biologists, environmentalists, political leaders and concerned citizens.

A similar term in the United States is *"natural heritage."* It pre-dates the others and is more accepted by the wider audience interested in conservation. Broader than biodiversity, it includes geology and landforms.

Definitions

A sampling of fungi collected during summer 2008 in Northern Saskatchewan mixed woods, near LaRonge is an example regarding the species diversity of fungi. In this photo, there are also leaf lichens and mosses.

"Biodiversity" is most commonly used to replace the more clearly defined and long established terms, species diversity and species richness. Biologists most often define biodiversity as the "totality of genes, species and ecosystems of a region". An advantage of this definition is that it seems to describe most circumstances and presents a unified view of the traditional types of biological variety previously identified:

- taxonomic diversity (usually measured at the species diversity level)
- ecological diversity often viewed from the perspective of ecosystem diversity
- morphological diversity which stems from genetic diversity

- functional diversity which is a measure of the number of functionally disparate species within a population (e.g. different feeding mechanism, different motility, predator vs prey, etc.)

In 2003 Professor Anthony Campbell at Cardiff University, UK and the Darwin Centre, Pembrokeshire, defined a fourth level: Molecular Diversity.

This multilevel construct is consistent with Datman and Lovejoy. An explicit definition consistent with this interpretation was first given in a paper by Bruce A. Wilcox commissioned by the International Union for the Conservation of Nature and Natural Resources (IUCN) for the 1982 World National Parks Conference. Wilcox's definition was "Biological diversity is the variety of life forms...at all levels of biological systems (i.e., molecular, organismic, population, species and ecosystem)...". The 1992 United Nations Earth Summit defined "biological diversity" as "the variability among living organisms from all sources, including, 'inter alia', terrestrial, marine and other aquatic ecosystems and the ecological complexes of which they are part: this includes diversity within species, between species and of ecosystems". This definition is used in the United Nations Convention on Biological Diversity.

One textbook's definition is "variation of life at all levels of biological organization".

Genetically biodiversity can be defined as the diversity of alleles, genes and organisms. They study processes such as mutation and gene transfer that drive evolution.

Measuring diversity at one level in a group of organisms may not precisely correspond to diversity at other levels. However, tetrapod (terrestrial vertebrates) taxonomic and ecological diversity shows a very close correlation.

Distribution

A conifer forest in the Swiss Alps (National Park)

Biodiversity is not evenly distributed, rather it varies greatly across the globe as well as within regions. Among other factors, the diversity of all living things (biota) depends on temperature, precipitation, altitude, soils, geography and the presence of other species. The study of the spatial distribution of organisms, species and ecosystems, is the science of biogeography.

Diversity consistently measures higher in the tropics and in other localized regions such as the Cape Floristic Region and lower in polar regions generally. Rain forests that have had wet climates for a long time, such as Yasuní National Park in Ecuador, have particularly high biodiversity.

Terrestrial biodiversity is thought to be up to 25 times greater than ocean biodiversity. A recently discovered method put the total number of species on Earth at 8.7 million, of which 2.1 million were estimated to live in the ocean. However, this estimate seems to under-represent the diversity of microorganisms.

Latitudinal Gradients

Generally, there is an increase in biodiversity from the poles to the tropics. Thus localities at lower latitudes have more species than localities at higher latitudes. This is often referred to as the latitudinal gradient in species diversity. Several ecological mechanisms may contribute to the gradient, but the ultimate factor behind many of them is the greater mean temperature at the equator compared to that of the poles.

Even though terrestrial biodiversity declines from the equator to the poles, some studies claim that this characteristic is unverified in aquatic ecosystems, especially in marine ecosystems. The latitudinal distribution of parasites does not appear to follow this rule.

Hotspots

A biodiversity hotspot is a region with a high level of endemic species that has experienced great habitat loss. The term hotspot was introduced in 1988 by Norman Myers. While hotspots are spread all over the world, the majority are forest areas and most are located in the tropics.

Brazil's Atlantic Forest is considered one such hotspot, containing roughly 20,000 plant species, 1,350 vertebrates and millions of insects, about half of which occur nowhere else. The island of Madagascar and India are also particularly notable. Colombia is characterized by high biodiversity, with the highest rate of species by area unit worldwide and it has the largest number of endemics (species that are not found naturally anywhere else) of any country. About 10% of the species of the Earth can be found in Colombia, including over 1,900 species of bird, more than in Europe and North America combined, Colombia has 10% of the world's mammals species, 14% of the amphibian species and 18% of the bird species of the world. Madagascar dry deciduous forests and lowland rainforests possess a high ratio of endemism. Since the island separated from mainland Africa 66 million years ago, many species and ecosystems have evolved independently. Indonesia's 17,000 islands cover 735,355 square miles (1,904,560 km^2) and contain 10% of the world's flowering plants, 12% of mammals and 17% of reptiles, amphibians and birds—along with nearly 240 million people. Many regions of high biodiversity and/or endemism arise from specialized habitats which require unusual adaptations, for example, alpine environments in high mountains, or Northern European peat bogs.

Accurately measuring differences in biodiversity can be difficult. Selection bias amongst researchers may contribute to biased empirical research for modern estimates of biodiversity. In 1768, Rev. Gilbert White succinctly observed of his Selborne, Hampshire *"all nature is so full, that that district produces the most variety which is the most examined."*

Evolution and History

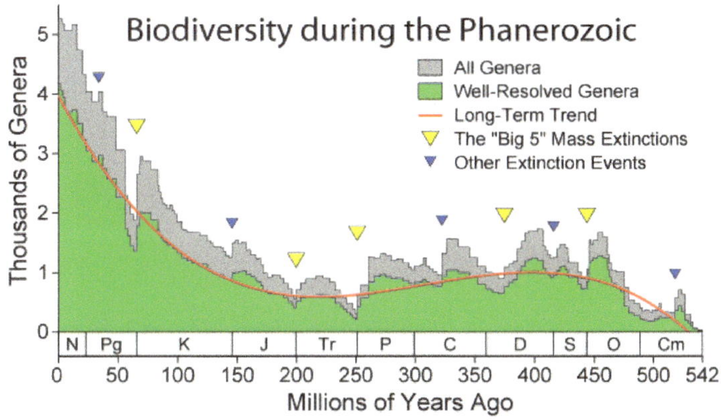

Apparent marine fossil diversity during the Phanerozoic

Biodiversity is the result of 3.5 billion years of evolution. The origin of life has not been definitely established by science, however some evidence suggests that life may already have been well-established only a few hundred million years after the formation of the Earth. Until approximately 600 million years ago, all life consisted of archaea, bacteria, protozoans and similar single-celled organisms.

The history of biodiversity during the Phanerozoic (the last 540 million years), starts with rapid growth during the Cambrian explosion—a period during which nearly every phylum of multicellular organisms first appeared. Over the next 400 million years or so, invertebrate diversity showed little overall trend and vertebrate diversity shows an overall exponential trend. This dramatic rise in diversity was marked by periodic, massive losses of diversity classified as mass extinction events. A significant loss occurred when rainforests collapsed in the carboniferous. The worst was the Permian-Triassic extinction event, 251 million years ago. Vertebrates took 30 million years to recover from this event.

The fossil record suggests that the last few million years featured the greatest biodiversity in history. However, not all scientists support this view, since there is uncertainty as to how strongly the fossil record is biased by the greater availability and preservation of recent geologic sections. Some scientists believe that corrected for sampling artifacts, modern biodiversity may not be much different from biodiversity 300 million years ago., whereas others consider the fossil record reasonably reflective of the diversification of life. Estimates of the present global macroscopic species diversity vary from 2 million to 100 million, with a best estimate of somewhere near 9 million, the vast majority arthropods. Diversity appears to increase continually in the absence of natural selection.

Evolutionary Diversification

The existence of a "global carrying capacity", limiting the amount of life that can live at once, is debated, as is the question of whether such a limit would also cap the number of species. While records of life in the sea shows a logistic pattern of growth, life on land (insects, plants and tetrapods)shows an exponential rise in diversity. As one author states, "Tetrapods have not yet invaded 64 per cent of potentially habitable modes and it could be that without human influence the ecological and taxonomic diversity of tetrapods would continue to increase in an exponential fashion until most or all of the available ecospace is filled."

On the other hand, changes through the Phanerozoic correlate much better with the hyperbolic model (widely used in population biology, demography and macrosociology, as well as fossil biodiversity) than with exponential and logistic models. The latter models imply that changes in diversity are guided by a first-order positive feedback (more ancestors, more descendants) and/or a negative feedback arising from resource limitation. Hyperbolic model implies a second-order positive feedback. The hyperbolic pattern of the world population growth arises from a second-order positive feedback between the population size and the rate of technological growth. The hyperbolic character of biodiversity growth can be similarly accounted for by a feedback between diversity and community structure complexity. The similarity between the curves of biodiversity and human population probably comes from the fact that both are derived from the interference of the hyperbolic trend with cyclical and stochastic dynamics.

Most biologists agree however that the period since human emergence is part of a new mass extinction, named the Holocene extinction event, caused primarily by the impact humans are having on the environment. It has been argued that the present rate of extinction is sufficient to eliminate most species on the planet Earth within 100 years.

New species are regularly discovered (on average between 5–10,000 new species each year, most of them insects) and many, though discovered, are not yet classified (estimates are that nearly 90% of all arthropods are not yet classified). Most of the terrestrial diversity is found in tropical forests and in general, land has more species than the ocean; some 8.7 million species may exists on Earth, of which some 2.1 million live in the ocean

Biodiversity and Ecosystem Services

Summer field in Belgium (Hamois). The blue flowers are *Centaurea cyanus* and the red are *Papaver rhoeas*.

The Balance of Evidence

"Ecosystem services are the suite of benefits that ecosystems provide to humanity."

These services come in three flavors:

1. Provisioning services which involve the production of renewable resources (e.g.: food, wood, fresh water)
2. Regulating services which are those that lessen environmental change (e.g.: climate regulation, pest/disease control)
3. Cultural services represent human value and enjoyment (e.g.: landscape aesthetics, cultural heritage, outdoor recreation and spiritual significance)

There have been many claims about biodiversity's effect on these ecosystem services, especially provisioning and regulating services. After an exhaustive survey through peer-reviewed literature to evaluate 36 different claims about biodiversity's effect on ecosystem services, 14 of those claims have been validated, 6 demonstrate mixed support or are unsupported, 3 are incorrect and 13 lack enough evidence to draw definitive conclusions.

Services Enhanced by Biodiversity

Provisioning Services

- Greater species diversity of plants increases fodder yield (synthesis of 271 experimental studies).
- Greater genetic diversity of plants (i.e.: diversity within a single species) in-

creases overall crop yield (synthesis of 575 experimental studies). Although another review of 100 experimental studies reports mixed evidence.

- Greater species diversity of trees increases overall wood production (Synthesis of 53 experimental studies). However, there is not enough data to draw a conclusion about the effect of tree trait diversity on wood production.

Regulating Services

- Greater species diversity of fish increases the stability of fisheries yield (Synthesis of 8 observational studies)
- Greater species diversity of natural pest enemies decreases herbivorous pest populations (Data from two separate reviews; Synthesis of 266 experimental and observational studies; Synthesis of 18 observational studies. Although another review of 38 experimental studies found mixed support for this claim, suggesting that in cases where mutual intraguild predation occurs, a single predatory species is often more effective
- Greater species diversity of plants decreases disease prevalence on plants (Synthesis of 107 experimental studies)
- Greater species diversity of plants increases resistance to plant invasion (Data from two separate reviews; Synthesis of 105 experimental studies; Synthesis of 15 experimental studies)
- Greater species diversity of plants increases carbon sequestration, but note that this finding only relates to actual uptake of carbon dioxide and not long term storage
- Greater species diversity of plants increases soil nutrient remineralization (Synthesis of 103 experimental studies)
- Greater species diversity of plants increases soil organic matter (Synthesis of 85 experimental studies)

Services with Mixed Evidence

Provisioning Services

- None to date

Regulating Services

- Greater species diversity of plants may or may not decrease herbivorous pest populations. Data from two separate reviews suggest that greater diversity decreases pest populations (Synthesis of 40 observational studies; Synthesis of

100 experimental studies). One review found mixed evidence (Synthesis of 287 experimental studies), while another found contrary evidence (Synthesis of 100 experimental studies)

- Greater species diversity of animals may or may not decrease disease prevalence on those animals (Synthesis of 45 experimental and observational studies), although a 2013 study offers more support showing that biodiversity may in fact enhance disease resistance within animal communities, at least in amphibian frog ponds. Many more studies must be published in support of diversity to sway the balance of evidence will be such that we can draw a general rule on this service.
- Greater species and trait diversity of plants may or may not increase long term carbon storage (Synthesis of 33 observational studies)
- Greater pollinator diversity may or may not increase pollination (Synthesis of 7 observational studies), but a publication from March 2013 suggests that increased native pollinator diversity enhances pollen deposition (although not necessarily fruit set as the authors would have you believe, for details explore their lengthy supplementary material).

Services for Which Biodiversity is a Hindrance

Provisioning Services

- Greater species diversity of plants reduces primary production (Synthesis of 7 experimental studies)

Regulating Services

- Greater genetic and species diversity of a number of organisms reduces freshwater purification (Synthesis of 8 experimental studies, although an attempt by the authors to investigate the effect of detritivore diversity on freshwater purification was unsuccessful due to a lack of available evidence (only 1 observational study was found

Provisioning Services

- Effect of species diversity of plants on biofuel yield (In a survey of the literature, the investigators only found 3 studies)
- Effect of species diversity of fish on fishery yield (In a survey of the literature, the investigators only found 4 experimental studies and 1 observational study)

Regulating Services

- Effect of species diversity on the stability of biofuel yield (In a survey of the literature, the investigators did not find any studies)

- Effect of species diversity of plants on the stability of fodder yield (In a survey of the literature, the investigators only found 2 studies)
- Effect of species diversity of plants on the stability of crop yield (In a survey of the literature, the investigators only found 1 study)
- Effect of genetic diversity of plants on the stability of crop yield (In a survey of the literature, the investigators only found 2 studies)
- Effect of diversity on the stability of wood production (In a survey of the literature, the investigators could not find any studies)
- Effect of species diversity of multiple taxa on erosion control (In a survey of the literature, the investigators could not find any studies – they did however find studies on the effect of species diversity and root biomass)
- Effect of diversity on flood regulation (In a survey of the literature, the investigators could not find any studies)
- Effect of species and trait diversity of plants on soil moisture (In a survey of the literature, the investigators only found 2 studies)

Other sources have reported somewhat conflicting results and in 1997 Robert Costanza and colleagues reported the estimated global value of ecosystem services (not captured in traditional markets) at an average of $33 trillion annually.

Since the stone age, species loss has accelerated above the average basal rate, driven by human activity. Estimates of species losses are at a rate 100-10,000 times as fast as is typical in the fossil record. Biodiversity also affords many non-material benefits including spiritual and aesthetic values, knowledge systems and education.

Biodiversity and Agriculture

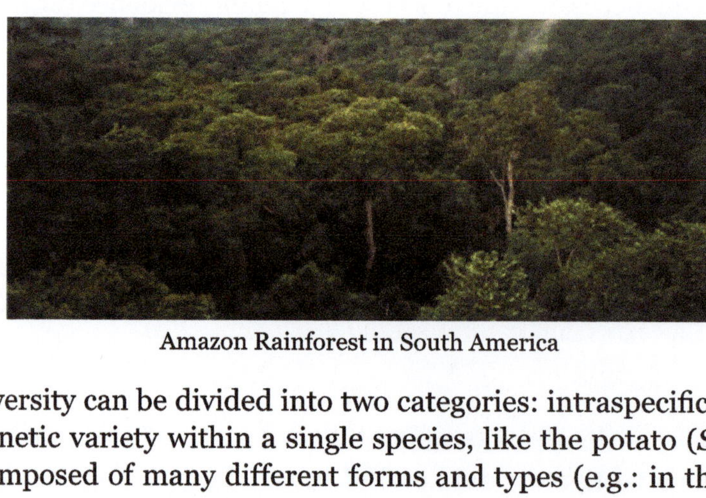

Amazon Rainforest in South America

Agricultural diversity can be divided into two categories: intraspecific diversity, which includes the genetic variety within a single species, like the potato (*Solanum tuberosum*) that is composed of many different forms and types (e.g.: in the U.S. we might compare russet potatoes with new potatoes or purple potatoes, all different, but all part of the same species, *S. tuberosum*).

The other category of agricultural diversity is called interspecific diversity and refers to the number and types of different species. Thinking about this diversity we might note that many small vegetable farmers grow many different crops like potatoes and also carrots, peppers, lettuce etc.

Agricultural diversity can also be divided by whether it is 'planned' diversity or 'associated' diversity. This is a functional classification that we impose and not an intrinsic feature of life or diversity. Planned diversity includes the crops which a farmer has encouraged, planted or raised (e.g.: crops, covers, symbionts and livestock, among others), which can be contrasted with the associated diversity that arrives among the crops, uninvited (e.g.: herbivores, weed species and pathogens, among others).

The control of associated biodiversity is one of the great agricultural challenges that farmers face. On monoculture farms, the approach is generally to eradicate associated diversity using a suite of biologically destructive pesticides, mechanized tools and transgenic engineering techniques, then to rotate crops. Although some polyculture farmers use the same techniques, they also employ integrated pest management strategies as well as strategies that are more labor-intensive, but generally less dependent on capital, biotechnology and energy.

Interspecific crop diversity is, in part, responsible for offering variety in what we eat. Intraspecific diversity, the variety of alleles within a single species, also offers us choice in our diets. If a crop fails in a monoculture, we rely on agricultural diversity to replant the land with something new. If a wheat crop is destroyed by a pest we may plant a hardier variety of wheat the next year, relying on intraspecific diversity. We may forgo wheat production in that area and plant a different species altogether, relying on interspecific diversity. Even an agricultural society which primarily grows monocultures, relies on biodiversity at some point.

- The Irish potato blight of 1846 was a major factor in the deaths of one million people and the emigration of about two million. It was the result of planting only two potato varieties, both vulnerable to the blight, *Phytophthora infestans*, which arrived in 1845

- When rice grassy stunt virus struck rice fields from Indonesia to India in the 1970s, 6,273 varieties were tested for resistance. Only one was resistant, an Indian variety and known to science only since 1966. This variety formed a hybrid with other varieties and is now widely grown.

- Coffee rust attacked coffee plantations in Sri Lanka, Brazil and Central America in 1970. A resistant variety was found in Ethiopia. The diseases are themselves a form of biodiversity.

Monoculture was a contributing factor to several agricultural disasters, including the European wine industry collapse in the late 19th century and the US southern corn leaf blight epidemic of 1970.

Although about 80 percent of humans' food supply comes from just 20 kinds of plants, humans use at least 40,000 species. Many people depend on these species for food, shelter and clothing. Earth's surviving biodiversity provides resources for increasing the range of food and other products suitable for human use, although the present extinction rate shrinks that potential.

Biodiversity and Human Health

The diverse forest canopy on Barro Colorado Island, Panama, yielded this display of different fruit

Biodiversity's relevance to human health is becoming an international political issue, as scientific evidence builds on the global health implications of biodiversity loss. This issue is closely linked with the issue of climate change, as many of the anticipated health risks of climate change are associated with changes in biodiversity (e.g. changes in populations and distribution of disease vectors, scarcity of fresh water, impacts on agricultural biodiversity and food resources etc.) This is because the species most likely to disappear are those that buffer against infectious disease transmission, while surviving species tend to be the ones that increase disease transmission, such as that of West Nile Virus, Lyme disease and Hantavirus, according to a study done co-authored by Felicia Keesing, an ecologist at Bard College and Drew Harvell, associate director for Environment of the Atkinson Center for a Sustainable Future (ACSF) at Cornell University.

The growing demand and lack of drinkable water on the planet presents an additional challenge to the future of human health. Partly, the problem lies in the success of water suppliers to increase supplies and failure of groups promoting preservation of water resources. While the distribution of clean water increases, in some parts of the world it remains unequal. According to *2008 World Population Data Sheet*, only 62% of least developed countries are able to access clean water.

Some of the health issues influenced by biodiversity include dietary health and nutrition security, infectious disease, medical science and medicinal resources, social and psychological health. Biodiversity is also known to have an important role in reducing disaster risk and in post-disaster relief and recovery efforts.

Biodiversity provides critical support for drug discovery and the availability of medicinal resources. A significant proportion of drugs are derived, directly or indirectly, from biological sources: at least 50% of the pharmaceutical compounds on the US market are derived from plants, animals and micro-organisms, while about 80% of the world population depends on medicines from nature (used in either modern or traditional medical practice) for primary healthcare. Only a tiny fraction of wild species has been investigated for medical potential. Biodiversity has been critical to advances throughout the field of bionics. Evidence from market analysis and biodiversity science indicates that the decline in output from the pharmaceutical sector since the mid-1980s can be attributed to a move away from natural product exploration ("bioprospecting") in favor of genomics and synthetic chemistry, indeed claims about the value of undiscovered pharmaceuticals may not provide enough incentive for companies in free markets to search for them because of the high cost of development; meanwhile, natural products have a long history of supporting significant economic and health innovation. Marine ecosystems are particularly important, although inappropriate bioprospecting can increase biodiversity loss, as well as violating the laws of the communities and states from which the resources are taken.

Biodiversity, Business and Industry

Agriculture production, pictured is a tractor and a chaser bin

Many industrial materials derive directly from biological sources. These include building materials, fibers, dyes, rubber and oil. Biodiversity is also important to the security of resources such as water, timber, paper, fiber and food. As a result, biodiversity loss is a significant risk factor in business development and a threat to long term economic sustainability.

Biodiversity, Leisure, Cultural and Aesthetic Value

Biodiversity enriches leisure activities such as hiking, birdwatching or natural history study. Biodiversity inspires musicians, painters, sculptors, writers and other artists. Many cultures view themselves as an integral part of the natural world which requires them to respect other living organisms.

Popular activities such as gardening, fishkeeping and specimen collecting strongly depend on biodiversity. The number of species involved in such pursuits is in the tens of thousands, though the majority do not enter commerce.

The relationships between the original natural areas of these often exotic animals and plants and commercial collectors, suppliers, breeders, propagators and those who promote their understanding and enjoyment are complex and poorly understood. The general public responds well to exposure to rare and unusual organisms, reflecting their inherent value.

Philosophically it could be argued that biodiversity has intrinsic aesthetic and spiritual value to mankind *in and of itself*. This idea can be used as a counterweight to the notion that tropical forests and other ecological realms are only worthy of conservation because of the services they provide.

Biodiversity and Ecological Services

Eagle Creek, Oregon hiking

Biodiversity supports many ecosystem services:

"There is now unequivocal evidence that biodiversity loss reduces the efficiency by which ecological communities capture biologically essential resources, produce biomass, decompose and recycle biologically essential nutrients... There is mounting evidence that biodiversity increases the stability of ecosystem functions through time... Diverse communities are more productive because they contain key species that have a large influence on productivity and differences in functional traits among organisms increase total resource capture... The impacts of diversity loss on ecological processes might be sufficiently large to rival the impacts of many other global drivers of environmental change... Maintaining multiple ecosystem processes at multiple places and times requires higher levels of biodiversity than does a single process at a single place and time."

It plays a part in regulating the chemistry of our atmosphere and water supply. Biodiversity is directly involved in water purification, recycling nutrients and providing fertile soils. Experiments with controlled environments have shown that humans cannot easily build ecosystems to support human needs; for example insect pollination cannot be mimicked, and that activity alone represented between $2.1-14.6 billions in 2003.

Number of Species

Species	Earth			Ocean		
	Catalogued	Predicted	±SE	Catalogued	Predicted	±SE
Eukaryotes						
Animalia	953,434	7,770,000	958,000	171,082	2,150,000	145,000
Chromista	13,033	27,500	30,500	4,859	7,400	9,640
Fungi	43,271	611,000	297,000	1,097	5,320	11,100
Plantae	215,644	298,000	8,200	8,600	16,600	9,130
Protozoa	8,118	36,400	6,690	8,118	36,400	6,690
Total	1,233,500	8,740,000	1,300,000	193,756	2,210,000	182,000
Prokaryotes						
Archaea	502	455	160	1	1	0
Bacteria	10,358	9,680	3,470	652	1,320	436
Total	10,860	10,100	3,630	653	1,320	436
Grand Total	**1,244,360**	**8,750,000**	**1,300,000**	**194,409**	**2,210,000**	**182,000**

Predictions for prokaryotes represent a lower bound because they do not consider undescribed higher taxa. For protozoa, the ocean database was substantially more complete than the database for the entire Earth so we only used the former to estimate the total number of species in this taxon. All predictions were rounded to three significant digits.

doi:10.1371/journal.pbio.1001127.t002

Discovered and predicted total number of species on land and in the oceans

According to Mora and colleagues, the total number of terrestrial species is estimated to be around 8.7 million while the number of oceanic species is much lower, estimated at 2.2 million. The authors note that these estimates are strongest for eukaryotic organisms and likely represent the lower bound of prokaryote diversity. Other estimates include:

- 220,000 vascular plants, estimated using the species-area relation method
- 0.7-1 million marine species
- 10–30 million insects; (of some 0.9 million we know today)
- 5–10 million bacteria;
- 1.5-3 million fungi, estimates based on data from the tropics, long-term non-tropical sites and molecular studies that have revealed cryptic speciation. Some 0.075 million species of fungi had been documented by 2001)
- 1 million mites

- The number of microbial species is not reliably known, but the Global Ocean Sampling Expedition dramatically increased the estimates of genetic diversity by identifying an enormous number of new genes from near-surface plankton samples at various marine locations, initially over the 2004-2006 period. The findings may eventually cause a significant change in the way science defines species and other taxonomic categories.

Since the rate of extinction has increased, many extant species may become extinct before they are described.

Species Loss Rates

No longer do we have to justify the existence of humid tropical forests on the feeble grounds that they might carry plants with drugs that cure human disease. *Gaia theory* forces us to see that they offer much more than this. Through their capacity to evapotranspirate vast volumes of water vapor, they serve to keep the planet cool by wearing a sunshade of white reflecting cloud. Their replacement by cropland could precipitate a disaster that is global in scale.

During the last century, decreases in biodiversity have been increasingly observed. In 2007, German Federal Environment Minister Sigmar Gabriel cited estimates that up to 30% of all species will be extinct by 2050. Of these, about one eighth of known plant species are threatened with extinction. Estimates reach as high as 140,000 species per year (based on Species-area theory). This figure indicates unsustainable ecological practices, because few species emerge each year. Almost all scientists acknowledge that the rate of species loss is greater now than at any time in human history, with extinctions occurring at rates hundreds of times higher than background extinction rates. As of 2012, some studies suggest that 25% of all mammal species could be extinct in 20 years.

In absolute terms, the planet has lost 52% of its biodiversity since 1970 according to a 2014 study by the World Wildlife Fund. The Living Planet Report 2014 claims that "the number of mammals, birds, reptiles, amphibians and fish across the globe is, on average, about half the size it was 40 years ago". Of that number, 39% accounts for the terrestrial wildlife gone, 39% for the marine wildlife gone and 76% for the freshwater wildlife gone. Biodiversity took the biggest hit in Latin America, plummeting 83 percent. High-income countries showed a 10% increase in biodiversity, which was canceled out by a loss in low-income countries. This is despite the fact that high-income countries use five times the ecological resources of low-income countries, which was explained as a result of process whereby wealthy nations are outsourcing resource depletion to poorer nations, which are suffering the greatest ecosystem losses.

Threats

In 2006 many species were formally classified as rare or endangered or threatened;

moreover, scientists have estimated that millions more species are at risk which have not been formally recognized. About 40 percent of the 40,177 species assessed using the IUCN Red List criteria are now listed as threatened with extinction—a total of 16,119.

Jared Diamond describes an "Evil Quartet" of habitat destruction, overkill, introduced species and secondary extinctions. Edward O. Wilson prefers the acronym HIPPO, standing for Habitat destruction, Invasive species, Pollution, human over-Population and Over-harvesting. The most authoritative classification in use today is IUCN's Classification of Direct Threats which has been adopted by major international conservation organizations such as the US Nature Conservancy, the World Wildlife Fund, Conservation International and BirdLife International.

Habitat Destruction

Deforestation and increased road-building in the Amazon Rainforest are a significant concern because of increased human encroachment upon wild areas, increased resource extraction and further threats to biodiversity.

Habitat destruction has played a key role in extinctions, especially related to tropical forest destruction. Factors contributing to habitat loss are: overconsumption, overpopulation, land use change, deforestation, pollution (air pollution, water pollution, soil contamination) and global warming or climate change.

Habitat size and numbers of species are systematically related. Physically larger species and those living at lower latitudes or in forests or oceans are more sensitive to reduction in habitat area. Conversion to "trivial" standardized ecosystems (e.g., monoculture following deforestation) effectively destroys habitat for the more diverse species that preceded the conversion. In some countries lack of property rights or lax law/regulatory enforcement necessarily leads to biodiversity loss (degradation costs having to be supported by the community).

A 2007 study conducted by the National Science Foundation found that biodiversity and genetic diversity are codependent—that diversity among species requires diversity within a species and vice versa. "If any one type is removed from the system, the cycle can break down and the community becomes dominated by a single species." At present, the most threatened ecosystems are found in fresh water, according to the Millennium Ecosystem Assessment 2005, which was confirmed by the "Freshwater Animal Diversity Assessment", organised by the biodiversity platform and the French *Institut de recherche pour le développement* (MNHNP).

Co-extinctions are a form of habitat destruction. Co-extinction occurs when the extinction or decline in one accompanies the other, such as in plants and beetles.

Introduced and Invasive Species

Male *Lophura nycthemera* (silver pheasant), a native of East Asia that has been introduced into parts of Europe for ornamental reasons

Barriers such as large rivers, seas, oceans, mountains and deserts encourage diversity by enabling independent evolution on either side of the barrier, via the process of allopatric speciation. The term invasive species is applied to species that breach the natural barriers that would normally keep them constrained. Without barriers, such species occupy new territory, often supplanting native species by occupying their niches, or by using resources that would normally sustain native species.

The number of species invasions has been on the rise at least since the beginning of the 1900s. Species are increasingly being moved by humans (on purpose and accidentally). In some cases the invaders are causing drastic changes and damage to their new habitats (e.g.: zebra mussels and the emerald ash borer in the Great Lakes region and the lion fish along the North American Atlantic coast). Some evidence suggests that invasive species are competitive in their new habitats because they are subject to less pathogen disturbance. Others report confounding evidence that occasionally suggest that species-rich communities harbor many native and exotic species simultaneously while some say that diverse ecosystems are more resilient and resist invasive plants and animals. An important question is, "do invasive species cause extinctions?" Many studies cite effects of invasive species on natives, but not extinctions. Invasive species seem to increase local (i.e.: alpha diversity) diversity, which decreases turnover of diversity (i.e.: beta diversity). Overall gamma diversity may be lowered because species are going extinct because of other causes, but even some of the most insidious invaders (e.g.: Dutch elm disease, emerald ash borer, chestnut blight in North America) have not caused their host species to become extinct. Extirpation, population decline and homogenization of regional biodiversity are much more common. Human activities have frequently been the cause of invasive species circumventing their barriers, by introducing them for food and other purposes. Human activities therefore allow species to migrate to new areas (and thus become invasive) occurred on time scales much shorter than historically have been required for a species to extend its range.

Not all introduced species are invasive, nor all invasive species deliberately introduced. In cases such as the zebra mussel, invasion of US waterways was unintentional. In other cases, such as mongooses in Hawaii, the introduction is deliberate but ineffective (nocturnal rats were not vulnerable to the diurnal mongoose). In other cases, such as oil palms in Indonesia and Malaysia, the introduction produces substantial economic benefits, but the benefits are accompanied by costly unintended consequences.

Finally, an introduced species may unintentionally injure a species that depends on the species it replaces. In Belgium, Prunus spinosa from Eastern Europe leafs much sooner than its West European counterparts, disrupting the feeding habits of the *Thecla betulae* butterfly (which feeds on the leaves). Introducing new species often leaves endemic and other local species unable to compete with the exotic species and unable to survive. The exotic organisms may be predators, parasites, or may simply outcompete indigenous species for nutrients, water and light.

At present, several countries have already imported so many exotic species, particularly agricultural and ornamental plants, that their own indigenous fauna/flora may be outnumbered. For example, the introduction of kudzu from Southeast Asia to Canada and the United States has threatened biodiversity in certain areas.

Genetic Pollution

Endemic species can be threatened with extinction through the process of genetic pollution, i.e. uncontrolled hybridization, introgression and genetic swamping. Genetic pollution leads to homogenization or replacement of local genomes as a result of either a numerical and/or fitness advantage of an introduced species. Hybridization and introgression are side-effects of introduction and invasion. These phenomena can be especially detrimental to rare species that come into contact with more abundant ones. The abundant species can interbreed with the rare species, swamping its gene pool. This problem is not always apparent from morphological (outward appearance) observations alone. Some degree of gene flow is normal adaptation and not all gene and genotype constellations can be preserved. However, hybridization with or without introgression may, nevertheless, threaten a rare species' existence.

Overexploitation

Overexploitation occurs when a resource is consumed at an unsustainable rate. This occurs on land in the form of overhunting, excessive logging, poor soil conservation in agriculture and the illegal wildlife trade.

About 25% of world fisheries are now overfished to the point where their current biomass is less than the level that maximizes their sustainable yield.

The overkill hypothesis, a pattern of large animal extinctions connected with human migration patterns, can be used explain why megafaunal extinctions can occur within a relatively short time period.

Hybridization, Genetic Pollution/Erosion and Food Security

The Yecoro wheat (right) cultivar is sensitive to salinity, plants resulting from a hybrid cross with cultivar W4910 (left) show greater tolerance to high salinity

In agriculture and animal husbandry, the Green Revolution popularized the use of conventional hybridization to increase yield. Often hybridized breeds originated in developed countries and were further hybridized with local varieties in the developing world to create high yield strains resistant to local climate and diseases. Local governments and industry have been pushing hybridization. Formerly huge gene pools of various wild and indigenous breeds have collapsed causing widespread genetic erosion and genetic pollution. This has resulted in loss of genetic diversity and biodiversity as a whole.

(GM organisms) have genetic material altered by genetic engineering procedures such as recombinant DNA technology. GM crops have become a common source for genetic pollution, not only of wild varieties but also of domesticated varieties derived from classical hybridization.

Genetic erosion coupled with genetic pollution may be destroying unique genotypes, thereby creating a hidden crisis which could result in a severe threat to our food security. Diverse genetic material could cease to exist which would impact our ability to further hybridize food crops and livestock against more resistant diseases and climatic changes.

Climate Change

Polar bears on the sea ice of the Arctic Ocean, near the North Pole.
Climate change has started affecting bear populations.

Global warming is also considered to be a major potential threat to global biodiversity in the future. For example, coral reefs - which are biodiversity hotspots - will be lost within the century if global warming continues at the current trend.

Climate change has seen many claims about potential to affect biodiversity but evidence supporting the statement is tenuous. Increasing atmospheric carbon dioxide certainly affects plant morphology and is acidifying oceans, and temperature affects species ranges, phenology, and weather, but the major impacts that have been predicted are still just *potential* impacts. We have not documented major extinctions yet, even as climate change drastically alters the biology of many species.

In 2004, an international collaborative study on four continents estimated that 10 percent of species would become extinct by 2050 because of global warming. "We need to limit climate change or we wind up with a lot of species in trouble, possibly extinct," said Dr. Lee Hannah, a co-author of the paper and chief climate change biologist at the Center for Applied Biodiversity Science at Conservation International.

A recent study predicts that up to 35% of the world terrestrial carnivores and ungulates will be at higher risk of extinction by 2050 because of the joint effects of predicted climate and land-use change under business-as-usual human development scenarios.

Human Overpopulation

From 1950 to 2011, world population increased from 2.5 billion to 7 billion and is forecast to reach a plateau of more than 9 billion during the 21st century. Sir David King, former chief scientific adviser to the UK government, told a parliamentary inquiry: *"It is self-evident that the massive growth in the human population through the 20th century has had more impact on biodiversity than any other single factor."* At least until the middle of the 21st century, worldwide losses of pristine biodiverse land will probably depend much on the worldwide human birth rate.

According to a 2014 study by the World Wildlife Fund, the global human population already exceeds planet's biocapacity - it would take the equivalent of 1.5 Earths of bio-capacity to meet our current demands. The report further points that if everyone on the planet had the Footprint of the average resident of Qatar, we would need 4.8 Earths and if we lived the lifestyle of a typical resident of the USA, we would need 3.9 Earths.

The Holocene Extinction

Rates of decline in biodiversity in this sixth mass extinction match or exceed rates of loss in the five previous mass extinction events in the fossil record. Loss of biodiversity results in the loss of natural capital that supplies ecosystem goods and services. From the perspective of the method known as Natural Economy the economic value of 17 ecosystem services for Earth's biosphere (calculated in 1997) has an estimated value of US$33 trillion ($3.3x10^{13}$) per year.

Conservation

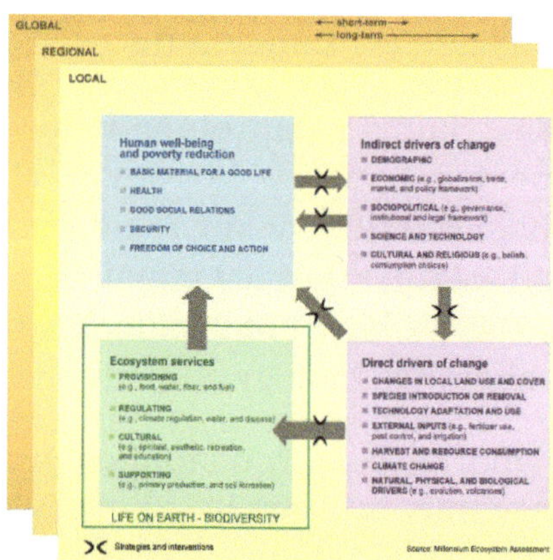

A schematic image illustrating the relationship between biodiversity, ecosystem services, human well-being and poverty. The illustration shows where conservation action, strategies and plans can influence the drivers of the current biodiversity crisis at local, regional, to global scales.

The retreat of Aletsch Glacier in the Swiss Alps (situation in 1979, 1991 and 2002), due to global warming.

Conservation biology matured in the mid-20th century as ecologists, naturalists and other scientists began to research and address issues pertaining to global biodiversity declines.

The conservation ethic advocates management of natural resources for the purpose of sustaining biodiversity in species, ecosystems, the evolutionary process and human culture and society.

Conservation biology is reforming around strategic plans to protect biodiversity. Preserving global biodiversity is a priority in strategic conservation plans that are designed to engage public policy and concerns affecting local, regional and global scales of communities, ecosystems and cultures. Action plans identify ways of sustaining human well-being, employing natural capital, market capital and ecosystem services.

In the EU Directive 1999/22/EC zoos are described as having a role in the preservation of the biodiversity of wildlife animals by conducting research or participation in breeding programs.

Protection and Restoration Techniques

Removal of exotic species will allow the species that they have negatively impacted to recover their ecological niches. Exotic species that have become pests can be identified taxonomically (e.g., with Digital Automated Identification SYstem (DAISY), using the barcode of life). Removal is practical only given large groups of individuals due to the economic cost.

As sustainable populations of the remaining native species in an area become assured, "missing" species that are candidates for reintroduction can be identified using databases such as the *Encyclopedia of Life* and the Global Biodiversity Information Facility.

- Biodiversity banking places a monetary value on biodiversity. One example is the Australian Native Vegetation Management Framework.
- Gene banks are collections of specimens and genetic material. Some banks intend to reintroduce banked species to the ecosystem (e.g., via tree nurseries).
- Reduction of and better targeting of pesticides allows more species to survive in agricultural and urbanized areas.
- Location-specific approaches may be less useful for protecting migratory species. One approach is to create wildlife corridors that correspond to the animals' movements. National and other boundaries can complicate corridor creation.

Protected Areas

Protected areas is meant for affording protection to wild animals and their habitat which also includes forest reserves and biosphere reserves. Protected areas have been set up all over the world with the specific aim of protecting and conserving plants and animals.

National Parks

National park and nature reserve is the area selected by governments or private organizations for special protection against damage or degradation with the objective of biodiversity and landscape conservation. National parks are usually owned and managed by national or state governments. A limit is placed on the number of visitors permitted to enter certain fragile areas. Designated trails or roads are created. The visitors are allowed to enter only for study, cultural and recreation purposes. Forestry operations, grazing of animals and hunting of animals are regulated. Exploitation of habitat or wildlife is banned.

Wildlife Sanctuary

Wildlife sanctuary aims only at conservation of species and have the following features:

1. The boundaries of the sanctuaries are not limited by state legislation.
2. The killing, hunting or capturing of any species is prohibited except by or under the control of the highest authority in the department which is responsible for the management of the sanctuary.
3. Private ownership may be allowed.
4. Forestry and other usages can also be permitted.

Forest Reserves

The forests play a vital role in harbouring more than 45,000 floral and 81,000 faunal species of which 5150 floral and 1837 faunal species are endemic. Plant and animal species confined to a specific geographical area are called endemic species. In reserved forests, rights to activities like hunting and grazing are sometimes given to communities living on the fringes of the forest, who sustain their livelihood partially or wholly from forest resources or products. The unclassed forests covers 6.4 percent of the total forest area and they are marked by the following characteristics:

1. They are large inaccessible forests.
2. Many of these are unoccupied.
3. They are ecologically and economically less important.

Steps to Conserve the Forest Cover

1. An extensive reforestation/afforestation program should be followed.
2. Alternative environment-friendly sources of fuel energy such as biogas other than wood should be used.
3. Loss of biodiversity due to forest fire is a major problem, immediate steps to prevent forest fire need to be taken.

4. Overgrazing by cattle can damage a forest seriously. Therefore, certain steps should be taken to prevent overgrazing by cattle.
5. Hunting and poaching should be banned.

Zoological Parks

In zoological parks or zoos, live animals are kept for public recreation, education and conservation purposes. Modern zoos offer veterinary facilities, provide opportunities for threatened species to breed in captivity and usually build environments that simulate the native habitats of the animals in their care. Zoos play a major role in creating awareness among common people about the need to conserve nature.

Botanical Gardens

Botanical garden is a garden in which plants are grown and displayed primarily for scientific and educational purposes. It consists of a collection of living plants, grown outdoors or under glass in greenhouses and conservatories. In addition, it includes a collection of dried plants or herbarium and such facilities as lecture rooms, laboratories, libraries, museums and experimental or research plantings.

Resource Allocation

Focusing on limited areas of higher potential biodiversity promises greater immediate return on investment than spreading resources evenly or focusing on areas of little diversity but greater interest in biodiversity.

A second strategy focuses on areas that retain most of their original diversity, which typically require little or no restoration. These are typically non-urbanized, non-agricultural areas. Tropical areas often fit both criteria, given their natively high diversity and relative lack of development.

Legal Status

A great deal of work is occurring to preserve the natural characteristics of Hopetoun Falls, Australia while continuing to allow visitor access.

International

- United Nations Convention on Biological Diversity (1992) and Cartagena Protocol on Biosafety;
- Convention on International Trade in Endangered Species (CITES);
- Ramsar Convention (Wetlands);
- Bonn Convention on Migratory Species;
- World Heritage Convention (indirectly by protecting biodiversity habitats)
- Regional Conventions such as the Apia Convention
- Bilateral agreements such as the Japan-Australia Migratory Bird Agreement.

Global agreements such as the Convention on Biological Diversity, give "sovereign national rights over biological resources" (not property). The agreements commit countries to "conserve biodiversity", "develop resources for sustainability" and "share the benefits" resulting from their use. Biodiverse countries that allow bioprospecting or collection of natural products, expect a share of the benefits rather than allowing the individual or institution that discovers/exploits the resource to capture them privately. Bioprospecting can become a type of biopiracy when such principles are not respected.

Sovereignty principles can rely upon what is better known as Access and Benefit Sharing Agreements (ABAs). The Convention on Biodiversity implies informed consent between the source country and the collector, to establish which resource will be used and for what and to settle on a fair agreement on benefit sharing.

National Level Laws

Biodiversity is taken into account in some political and judicial decisions:

- The relationship between law and ecosystems is very ancient and has consequences for biodiversity. It is related to private and public property rights. It can define protection for threatened ecosystems, but also some rights and duties (for example, fishing and hunting rights).
- Law regarding species is more recent. It defines species that must be protected because they may be threatened by extinction. The U.S. Endangered Species Act is an example of an attempt to address the "law and species" issue.
- Laws regarding gene pools are only about a century old. Domestication and plant breeding methods are not new, but advances in genetic engineering have led to tighter laws covering distribution of genetically modified organisms, gene patents and process patents. Governments struggle to decide whether to focus on for example, genes, genomes, or organisms and species.

Uniform approval for use of biodiversity as a legal standard has not been achieved, however. Bosselman argues that biodiversity should not be used as a legal standard, claiming that the remaining areas of scientific uncertainty cause unacceptable administrative waste and increase litigation without promoting preservation goals.

India passed the Biological Diversity Act in 2002 for the conservation of biological diversity in India. The Act also provides mechanisms for equitable sharing of benefits from the use of traditional biological resources and knowledge.

Analytical Limits

Taxonomic and Size Relationships

Less than 1% of all species that have been described have been studied beyond simply noting their existence. The vast majority of Earth's species are microbial. Contemporary biodiversity physics is "firmly fixated on the visible [macroscopic] world". For example, microbial life is metabolically and environmentally more diverse than multicellular life. "On the tree of life, based on analyses of small-subunit ribo-somal RNA, visible life consists of barely noticeable twigs. The inverse relationship of size and population recurs higher on the evolutionary ladder—"to a first approximation, all multicellular species on Earth are insects". Insect extinction rates are high—supporting the Holocene extinction hypothesis.

Environmental Protection

Environmental protection is a practice of protecting the natural environment on individual, organizational or governmental levels, for the benefit of both the environment and humans. Due to the pressures of overconsumption, population and technology, the biophysical environment is being degraded, sometimes permanently. This has been recognized, and governments have begun placing restraints on activities that cause environmental degradation. Since the 1960s, activity of environmental movements has created awareness of the various environmental issues. There is no agreement on the extent of the environmental impact of human activity and even scientific dishonesty occurs, so protection measures are occasionally debated.

Approaches

Voluntary Environmental Agreements

In industrial countries, voluntary environmental agreements often provide a platform for companies to be recognized for moving beyond the minimum regulatory standards and thus support the development of best environmental practice. In India Environment Improvement Trust (EIT) working for environment & forest protection since

1998. A group of Green Volunteers get a goal of Green India Clean India concept. CA Gajendra Kumar Jain an Chartered Accountant is founder of Environment Improvement Trust in Sojat city a small village of State of Rajasthan in India. In developing countries, such as throughout Latin America, these agreements are more commonly used to remedy significant levels of non-compliance with mandatory regulation. The challenges that exist with these agreements lie in establishing baseline data, targets, monitoring and reporting. Due to the difficulties inherent in evaluating effectiveness, their use is often questioned and, indeed, the whole environment may well be adversely affected as a result. The key advantage of their use in developing countries is that their use helps to build environmental management capacity.

Ecosystems Approach

An ecosystems approach to resource management and environmental protection aims to consider the complex interrelationships of an entire ecosystem in decision making rather than simply responding to specific issues and challenges. Ideally the decision-making processes under such an approach would be a collaborative approach to planning and decision making that involves a broad range of stakeholders across all relevant governmental departments, as well as representatives of industry, environmental groups and community. This approach ideally supports a better exchange of information, development of conflict-resolution strategies and improved regional conservation.

International Environmental Agreements

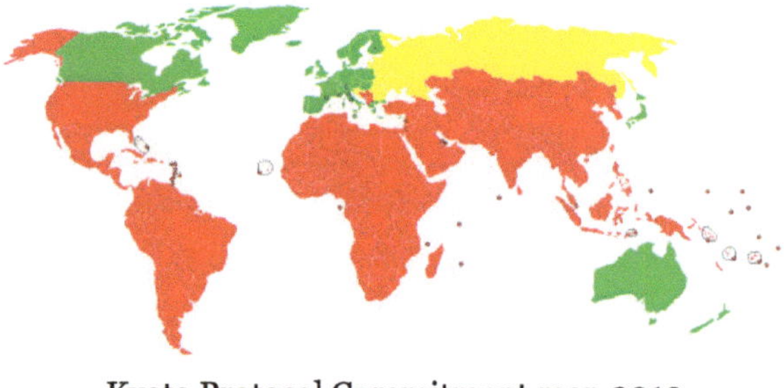

Kyoto Protocol Commitment map 2010

Many of the earth's resources are especially vulnerable because they are influenced by human impacts across many countries. As a result of this, many attempts are made by countries to develop agreements that are signed by multiple governments to prevent damage or manage the impacts of human activity on natural resources. This can include agreements that impact factors such as climate, oceans, rivers and air pollution. These international environmental agreements are sometimes legally binding documents that have legal implications when they are not followed and, at other times, are more agreements in principle or are for use as codes of conduct. These agreements have a long history with some multinational agreements being in place from as early as 1910 in Europe, America and Africa. Some of the most well-known multinational agreements include:

Government

Discussion concerning environmental protection often focuses on the role of government, legislation, and law enforcement. However, in its broadest sense, environmental protection may be seen to be the responsibility of all the people and not simply that of government. Decisions that impact the environment will ideally involve a broad range of stakeholders including industry, indigenous groups, environmental group and community representatives. Gradually, environmental decision-making processes are evolving to reflect this broad base of stakeholders and are becoming more collaborative in many countries.

Many constitutions acknowledge the fundamental right to environmental protection and many international treaties acknowledge the right to live in a healthy environment. Also, many countries have organizations and agencies devoted to environmental protection. There are international environmental protection organizations, such as the United Nations Environment Programme.

Although environmental protection is not simply the responsibility of government agencies, most people view these agencies as being of prime importance in establishing and maintaining basic standards that protect both the environment and the people interacting with it.

Tanzania

Zebras, Serengeti savana plains, Tanzania

Tanzania is recognised as having some of the greatest biodiversity of any African country. Almost 40% of the land has been established into a network of protected areas, including several national parks. The concerns for the natural environment include damage to ecosystems and loss of habitat resulting from population growth, expansion of subsistence agriculture, pollution, timber extraction and significant use of timber as fuel.

History of Environmental Protection

Environmental protection in Tanzania began during the German occupation of East Africa (1884-1919) — colonial conservation laws for the protection of game and forests were enacted, whereby restrictions were placed upon traditional indigenous activities such as hunting, firewood collecting and cattle grazing. In year 1948, Serengeti was officially established as the first national park for wild cats in East Africa. Since 1983, there has been a more broad-reaching effort to manage environmental issues at a national level, through the establishment of the National Environment Management Council (NEMC) and the development of an environmental act. In 1998 Environment Improvement Trust (EIT) start working for environment & forest protection in India from a small city Sojat. Founder of Environment Improvement Trust is CA Gajendra Kumar Jain working with volunteers.

Government Protection

Division of the biosphere is the main government body that oversees protection. It does this through the formulation of policy, coordinating and monitoring environmental issues, environmental planning and policy-oriented environmental research.The National Environment Management Council (NEMC) is an institution that was initiated when the National Environment Management Act was first introduced in year 1983. This council has the role to advise governments and the international community on a range of environmental issues. The NEMC the following purposes: provide technical advice; coordinate technical activities; develop enforcement guidelines and procedures; assess, monitor and evaluate activities that impact the environment; promote and assist environmental information and communication; and seek advancement of scientific knowledge.

The National Environment Policy of 1997 acts as a framework for environmental decision making in Tanzania. The policy objectives are to achieve the following:

- Ensure sustainable and equitable use of resources without degrading the environment or risking health or safety
- Prevent and control degradation of land, water, vegetation and air
- Conserve and enhance natural and man-made heritage, including biological diversity of unique ecosystems
- Improve condition and productivity of degraded areas
- Raise awareness and understanding of the link between environment and development
- Promote individual and community participation
- Promote international cooperation

Tanzania is a signatory to a significant number of international conventions including the Rio Declaration on Development and Environment 1992 and the Convention on Biological Diversity 1996. The Environmental Management Act, 2004, is the first comprehensive legal and institutional framework to guide environmental-management decisions. The policy tools that are parts of the act includes the use of: environmental-impact assessments, strategics environmentals assessments and taxation on pollution for specific industries and products. The effectiveness of shifing of this act will only become clear over time as concerns regarding its implementation become apparent based on the fact that, historically, there has been a lack of capacity to enforce environmental laws and a lack of working tools to bring environmental-protection objectives into practice.

China

The Longwanqun National Forest Park is a nationally protected nature area in Huinan County, Jilin, China

Formal environmental protection in China House was first stimulated by the 1972 United Nations Conference on the Human Environment held in Stockholm, Sweden. Following this, they began establishing environmental protection agencies and putting controls on some of its industrial waste. China was one of the first developing countries to implement a sustainable development strategy. In 1983 the State Council announced that environmental protection would be one of China's basic national policies and in 1984 the National Environmental Protection Agency (NEPA) was established. Following severe flooding of the Yangtze River basin in 1998, NEPA was upgraded to the State Environmental Protection Agency (SEPA) meaning that environmental protection was now being implemented at a ministerial level. In 2008, SEPA became known by its current name of Ministry of Environmental Protection of the People's Republic of China (MEP).

Pollution control instruments in China

Command-and-control	Economic incentives	Voluntary instruments	Public participation
Concentration-based pollution discharge controls	Pollution levy fee	Environmental labeling system	Clean-up campaign
Mass-based controls on total provincial discharge	Non-compliance fines	ISO 14000 system	Environmental awareness campaign

Environmental impact assessments (EIA)	Discharge permit system	Cleaner production	Air pollution index
Three synchronization program	Sulfur emission fee	NGOs	Water quality disclosure
Deadline transmission trading		Administrative permission hearing	
Centralized pollution control	Subsidies for energy saving products		
Two compliance policy	Regulation on refuse credit to high-polluting firms		
Environmental compensation fee			

Environmental pollution and ecological degradation has resulted in economic losses for China. In 2005, economic losses (mainly from air pollution) were calculated at 7.7% of China's GDP. This grew to 10.3% by 2002 and the economic loss from water pollution (6.1%) began to exceed that caused by air pollution. China has been one of the top performing countries in terms of GDP growth (9.64% in the past ten years). However, the high economic growth has put immense pressure on its environment and the environmental challenges that China faces are greater than most countries. In 2010 China was ranked 121st out of 163 countries on the Environmental Performance Index.

China has taken initiatives to increase its protection of the environment and combat environmental degradation:

- China's investment in renewable energy grew 18% in 2007 to $15.6 billion, accounting for ~10% of the global investment in this area;).
- In 2008, spending on the environment was 1.49% of GDP, up 3.4 times from 2000;
- The discharge of COD (carbon monoxide) and SO2 (sulfur dioxide) decreased by 6.61% and 8.95% in 2008 compared with that in 2005;
- China's protected nature reserves have increased substantially. In 1978 there were only 34 compared with 2,538 in 2010. The protected nature reserve system now occupies 15.5% of the country; this is higher than the world average.

Rapid growth in GDP has been China's main goal during the past three decades with a dominant development model of inefficient resource use and high pollution to achieve high GDP. For China to develop sustainably, environmental protection should be treated as an integral part of its economic policies.

Quote from Shengxian Zhou, head of MEP (2009): "Good economic policy is good environmental policy and the nature of environmental problem is the economic structure, production form and develop model."

European Union

Environmental protection has become an important task for the institutions of the European Community after the Maastricht Treaty for the European Union ratification by all the Member States. The EU is already very active in the field of environmental policy with important directives like those on environmental impact assessment and on the access to environmental information for citizens in the Member States.

Russia

In Russia, environmental protection is considered an integral part of national safety. There is an authorized state body - the Federal Ministry of Natural Resources and Ecology. However, there are a lot of environmental problems.

Latin America

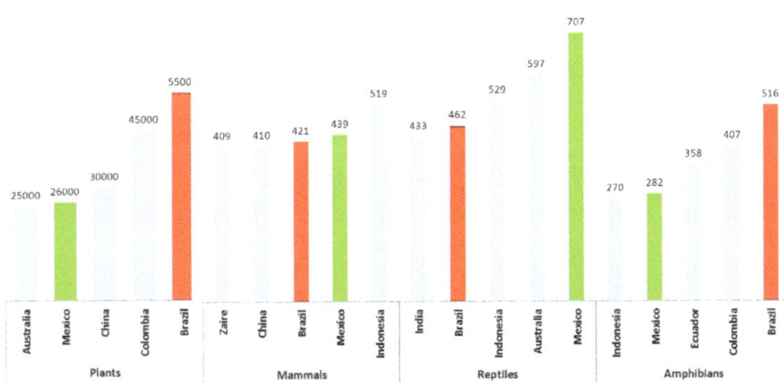

Top 5 Countries by biological diversity

The United Nations Environment Programme (UNEP) has identified 17 megadiverse countries. The list includes six Latin American countries: Brazil, Colombia, Ecuador, Mexico, Peru and Venezuela. Mexico and Brazil stand out among the rest because they have the largest area, population and number of species. These countries represent a major concern for environmental protection because they have high rates of deforestation, ecosystems loss, pollution, and population growth.

Brazil

Brazil has the largest amount of the world's tropical forests, 4,105,401 km2 (48.1% of Brazil), concentrated in the Amazon region. Brazil is home to vast biological diversity, first among the megadiverse countries of the world, having between 15%-20% of the 1.5 million globally described species.

Panorama of the Iguazu falls in Brazil

The organization in charge of environment protection is the Brazilian Ministry of the Environment (in Portuguese: Ministério do Meio Ambiente, MMA). It was first created in year 1973 with the name Special Secretariat for the Environment (Secretaria Especial de Meio Ambiente), changing names several times, and adopting the final name in year 1999. The Ministry is responsible for addressing the following issues:

- A national policy for the environment and for water resources;
- A policy for the preservation, conservation and sustainable use of ecosystems, biodiversity and forests;
- Proposing strategies, mechanisms, economic and social instruments for improving environmental quality, and sustainable use of natural resources;
- Policies for integrating production and the environment;
- Environmental policies and programs for the Legal Amazon;
- Ecological and economic territorial zoning.

In 2011, protected areas of the Amazon covered 2,197,485 km2 (an area larger than Greenland), with conservation units, like national parks, accounting for just over half (50.6%), and indigenous territories representing the remaining 49.4%.

Mexico

With over 200,000 different species, Mexico is home to 10–12% of the world's biodiversity, ranking first in reptile biodiversity and second in mammals—one estimate indicates that over 50% of all animal and plant species live in Mexico.

The history of environmental policy in Mexico started in the 1940s with the enactment of the Law of Conservation of Soil and Water (in Spanish: Ley de Conservación de Suelo y Agua). Three decades later, at the beginning of the 1970s, the Law to Prevent and

Control Environmental Pollution was created (Ley para Prevenir y Controlar la Contaminación Ambiental).

In year 1972 was the first direct response from the federal government to address eminent health effects from environmental issues. It established the administrative organization of the Secretariat for the Improvement of the Environment (Subsecretaría para el Mejoramiento del Ambiente) in the Department of Health and Welfare.

The axolotl is an endemic species from the central part of Mexico

The Secretariat of Environment and Natural Resources (Secretaría del Medio Ambiente y Recursos Naturales, SEMARNAT) is Mexico's environment ministry. The Ministry is responsible for addressing the following issues:

- Promote the protection, restoration and conservation of ecosystems, natural resources, goods and environmental services, and to facilitate their use and sustainable development.
- Develop and implement a national policy on natural resources
- Promote environmental management within the national territory, in coordination with all levels of government and the private sector.
- Evaluate and provide determination to the environmental impact statements for development projects and prevention of ecological damage
- Implement national policies on climate change and protection of the ozone layer.
- Direct work and studies on national meteorological, climatological, hydrological, and geohydrological systems, and participate in international conventions on these subjects.
- Regulate and monitor the conservation of waterways

In November 2000 there were 127 protected areas; currently there are 174, covering an area of 25,384,818 hectares, increasing federally protected areas from 8.6% to 12.85% its land area.

Oceania

Australia

The Great Barrier Reef in Australia is the largest barrier reef in the world

In 2008, there was 98,487,116 ha of terrestrial protected area, covering 12.8% of the land area of Australia. The 2002 figures of 10.1% of terrestrial area and 64,615,554 ha of protected marine area were found to poorly represent about half of Australia's 85 bioregions.

Environmental protection in Australia could be seen as starting with the formation of the first National Park, Royal National Park, in 1879. More progressive environmental protection had it start in the 1960s and 1970s with major international programs such as the United Nations Conference on the Human Environment in 1972, the Environment Committee of the OECD in 1970, and the United Nations Environment Programme of 1972. These events laid the foundations by increasing public awareness and support for regulation. State environmental legislation was irregular and deficient until the Australian Environment Council (AEC) and Council of Nature Conservation Ministers (CONCOM) were established in 1972 and 1974, creating a forum to assist in coordinating environmental and conservation policies between states and neighbouring countries. These councils have since been replaced by the Australian and New Zealand Environment and Conservation Council (ANZECC) in 1991 and finally the Environment Protection and Heritage Council (EPHC) in 2001.

At a national level, the Environment Protection and Biodiversity Conservation Act of 1999 is the primary environmental protection legislation for the Commonwealth of Australia. It concerns matters of national and international environmental significance regarding flora, fauna, ecological communities and cultural heritage. It also has jurisdiction over any activity conducted by the Commonwealth, or affecting it, that has significant environmental impact. The act covers eight main areas:

- National Heritage Sites
- World Heritage Sites
- RAMSAR wetlands
- Nationally endangered or threatened species and ecological communities
- Nuclear activities and actions
- The Great Barrier Reef Marine Park
- Migratory species
- Commonwealth Marine areas

There are several Commonwealth protected lands due to partnerships with traditional native owners, such as Kakadu National Park, extraordinary biodiversity such as Christmas Island National Park, or managed cooperatively due to cross-state location, such as the Australian Alps National parks.

At a state level, the bulk of environmental protection issues are left to the responsibility of the state or territory. Each state in Australia has its own environmental protection legislation and corresponding agencies. Their jurisdiction is similar and covers point-source pollution, such as from industry or commercial activities, land/water use, and waste management. Most protected lands are managed by states and territories with state legislative acts creating different degrees and definitions of protected areas such as wilderness, national land and marine parks, state forests, and conservation areas. States also create regulation to limit and provide general protection from air, water, and sound pollution.

At a local level, each city or regional council has responsibility over issues not covered by state or national legislation. This includes non-point source, or diffuse pollution, such as sediment pollution from construction sites.

Australia ranks second place on the UN 2010 Human Development Index and one of the lowest debt to GDP ratios of the developed economies. This could be seen as coming at the cost of the environment, with Australia being the world leader in coal exportation and species extinctions. Some have been motivated to proclaim it is Australia's responsibility to set the example of environmental reform for the rest of the world to follow.

New Zealand

At a national level, the Ministry for the Environment is responsible for environmental policy and the Department of Conservation addresses conservation issues. At a regional level the regional councils administer the legislation and address regional environmental issues.

United States

Yosemite National Park in California. One of the first protected areas in the United States

Since 1969, the United States Environmental Protection Agency (EPA) has been working to protect the environment and human health. All U.S. states have their own state departments of environmental protection.

The EPA has drafted "Seven Priorities for EPA's Future", which are:

- "Taking Action on Climate Change"
- "Improving Air Quality"
- "Assuring the Safety of Chemicals"
- "Cleaning Up Our Communities"
- "Protecting America's Waters"
- "Expanding the Conversation on Environmentalism and Working for Environmental Justice"
- "Building Strong State and Tribal Partnerships"

In Literature

There are many works of literature that contain the themes of environmental protection but some have been fundamental to its evolution. Several pieces such as *A Sand County Almanac* by Aldo Leopold, *Tragedy of the commons* by Garrett Hardin, and *Silent Spring* by Rachel Carson have become classics due to their far reaching influences. Environmental protection is present in fiction as well as non-fictional literature. Books such as *Antarctica* and *Blockade* have environmental protection as subjects whereas *The Lorax* has become a popular metaphor for environmental protection. "The Limits of Trooghaft" by Desmond Stewart is a short story that provides insight into human attitudes towards animals. Another book called "The Martian Chronicles" by Ray Bradbury investigates issues such as bombs, wars, government control, and what effects these can have on the environment.

Habitat Conservation

Tree planting is an aspect of habitat conservation. In each plastic tube a hardwood tree has been planted.

There are significant ecological benefits associated with selective cutting. Pictured is an area with Ponderosa Pine trees that were selectively harvested.

Habitat conservation is a management practice that seeks to conserve, protect and restore habitat areas for wild plants and animals, especially conservation reliant species, and prevent their extinction, fragmentation or reduction in range. It is a priority of many groups that cannot be easily characterized in terms of any one ideology.

History of the Conservation Movement

For much of human history, *nature* had been seen as a resource, one that could be controlled by the government and used for personal and economic gain. The idea was that plants only existed to feed animals and animals only existed to feed humans. The land itself had limited value only extending to the resources it could provide such as minerals and oil.

Throughout the 18th and 19th centuries social views started to change and scientific conservation principles were first practically applied to the forests of British India. The conservation ethic that began to evolve included three core principles: that human activity damaged the environment, that there was a civic duty to maintain the environment for future generations, and that scientific, empirically based methods should be applied to ensure this duty was carried out. Sir James Ranald Martin was prominent in promoting this ideology, publishing many medico-topographical reports that demonstrated the scale of damage wrought through large-scale deforestation and desiccation, and lobbying extensively for the institutionalization of forest conservation activities in British India through the establishment of Forest Departments.

The Madras Board of Revenue started local conservation efforts in 1842, headed by Alexander Gibson, a professional botanist who systematically adopted a forest conservation program based on scientific principles. This was the first case of state conservation management of forests in the world. Governor-General Lord Dalhousie introduced the first permanent and large-scale forest conservation program in the world in 1855, a model that soon spread to other colonies, as well the United States, where Yellowstone National Park was opened in 1872 as the world's first national park.

Rather than focusing on the economic or material benefits associated with nature, humans began to appreciate the value of nature itself and the need to protect pristine wilderness. By the middle of the 20th century countries such as the United States, Canada, and Britain understood this appreciation and instigated laws and legislation in order to ensure that the most fragile and beautiful environments would be protected for generations to come. Today with the help of NGO's, not-for profit organizations and governments world-wide there is a stronger movement taking place, with a deeper understanding of habitat conservation with the aim of protecting delicate habitats and preserving biodiversity on a global scale. The commitment and actions of small volunteering association in villages and towns, that endeavour to emulate the work done by well known Conservation Organisations, is paramount in ensuring generations that follow understand the importance of conserving natural resources. A village conservation group with the mission statement "We are committed to protecting and enhancing the natural environment in and around the adjoining villages of Ouston and Urpeth." may one day inspire a child who becomes the employee of a worldwide conservation organisation.

Values of Natural Habitat

The natural environment is a source for a wide range of resources that can be exploited for economic profit, for example timber is harvested from forests and clean water is obtained from natural streams. However, land development from anthropogenic economic growth often causes a decline in the ecological integrity of nearby natural habitat. For instance, this was an issue in the northern rocky mountains of the USA.

However, there is also economic value in conserving natural habitats. Financial profit can be made from tourist revenue, particularly in the tropics where species diversity is high. The cost of repairing damaged ecosystems is considered to be much higher than the cost of conserving natural ecosystems.

Measuring the worth of conserving different habitat areas is often criticized as being too utilitarian from a philosophical point of view.

Biodiversity

Habitat conservation is important in maintaining biodiversity, an essential part of global food security. There is evidence to support a trend of accelerating erosion of the genetic resources of agricultural plants and animals. An increase in genetic similarity of agricultural plants and animals means an increased risk of food loss from major epidemics. Wild species of agricultural plants have been found to be more resistant to disease, for example the wild corn species Teosinte is resistant to 4 corn diseases that affect human grown crops. A combination of seed banking and habitat conservation has been proposed to maintain plant diversity for food security purposes.

Classifying Environmental Values

Pearce and Moran outlined the following method for classifying environmental uses:

- Direct extractive uses: e.g. timber from forests, food from plants and animals
- Indirect uses: e.g. ecosystem services like flood control, pest control, erosion protection
- Optional uses: future possibilities e.g. unknown but potential use of plants in chemistry/medicine
- Non-use values:
    - Bequest value (benefit of an individual who knows that others may benefit from it in future)
    - Passive use value (sympathy for natural environment, enjoyment of the mere existence of a particular species)

Impacts

Natural Causes

Habitat loss and destruction can occur both naturally and through anthropogenic causes. Events leading to natural habitat loss include climate change, catastrophic events such as volcanic explosions and through the interactions of invasive and non-invasive species. Natural climate change, events have previously been the cause of many widespread and large scale losses in habitat. For example, some of the mass extinction events generally referred to as the "Big Five" have coincided with large scale such as the Earth entering an ice age, or alternate warming events. Other events in the big five also have their roots in natural causes, such as volcanic explosions and meteor collisions. The Chicxulub impact is one such example, which has previously caused widespread losses in habitat as the Earth either received less sunlight or grew colder, causing certain fauna and flora to flourish whilst others perished. Previously known warm areas in the tropics, the most sensitive habitats on Earth, grew colder, and areas such as Australia developed radically different flora and fauna to those seen today. The big five mass extinction events have also been linked to sea level changes, indicating that large scale marine species loss was strongly influenced by loss in marine habitats, particularly shelf habitats. Methane-driven oceanic eruptions have also been shown to have caused smaller mass extinction events.

Human Impacts

Humans have been the cause of many species' extinction. Due to humans' changing and modifying their environment, the habitat of other species often become altered or destroyed as a result of human actions. Even before the modern industrial era, humans were having widespread, and major effects on the environment. A good example of this is found in Aboriginal Australians and Australian megafauna. Aboriginal hunting practices, which included burning large sections of forest at a time, eventually altered and changed Australia's vegetation so much that many herbivorous megafauna species were left with no habitat and were driven into extinction. Once herbivorous megafauna species became extinct, carnivorous megafauna species soon followed. In the recent past, humans have been responsible for causing more extinctions within a given period of time than ever before. Deforestation, pollution, anthropogenic climate change and human settlements have all been driving forces in altering or destroying habitats. The destruction of ecosystems such as rainforests has resulted in countless habitats being destroyed. These biodiversity hotspots are home to millions of habitat specialists, which do not exist beyond a tiny area. Once their habitat is destroyed, they cease to exist.This destruction has a follow-on effect, as species which coexist or depend upon the existence of other species also become extinct, eventually resulting in the collapse of an entire ecosystem. These time-delayed extinctions are referred to as the extinction debt, which is the result of destroying and fragmenting habitats. As a result of anthropogenic modification of the environment, the extinction rate has climbed to the point where the Earth is now within a sixth mass extinction event, as commonly agreed by biologists. This has been particularly evident, for example, in the rapid decline in the number of amphibian species worldwide.

Approaches and Methods of Habitat Conservation

Determining the size, type and location of habitat to conserve is a complex area of conservation biology. Although difficult to measure and predict, the conservation value of a habitat is often a reflection of the quality (e.g. species abundance and diversity), endangerment of encompassing ecosystems, and spatial distribution of that habitat.

Identifying Priority Habitats for Conservation

Habitat conservation is vital for protecting species and ecological processes. It is important to conserve and protect the space/ area in which that species occupies. Therefore, areas classified as 'biodiversity hotspots', or those in which a flagship, umbrella, or endangered species inhabits are often the habitats that are given precedence over others. Species that possess an elevated risk of extinction are given the highest priority and as a result of conserving their habitat, other species in that community are protected thus serving as an element of gap analysis. In the United States of America, a Habitat Conservation Plan (HCP) is often developed to conserve the environment in which a specific species inhabits. Under the U.S. Endangered Species Act (ESA) the habitat that requires protection in an HCP is referred to as the 'critical habitat'. Multiple-species HCPs are becoming more favourable than single-species HCPs as they can potentially protect an array of species before they warrant listing under the ESA, as well as being able to conserve broad ecosystem components and processes . As of January 2007, 484 HCPs were permitted across the United States, 40 of which covered 10 or more species. The San Diego Multiple Species Conservation Plan (MSCP) encompasses 85 species in a total area of 26,000-km2. Its aim is to protect the habitats of multiple species and overall biodiversity by minimizing development in sensitive areas.

HCPs require clearly defined goals and objectives, efficient monitoring programs, as well as successful communication and collaboration with stakeholders and land owners in the area. Reserve design is also important and requires a high level of planning and management in order to achieve the goals of the HCP. Successful reserve design often takes the form of a hierarchical system with the most valued habitats requiring high protection being surrounded by buffer habitats that have a lower protection status. Like HCPs, hierarchical reserve design is a method most often used to protect a single species, and as a result habitat corridors are maintained, edge effects are reduced and a broader suite of species are protected.

How Much Habitat is Needed

A range of methods and models currently exist that can be used to determine how much habitat is to be conserved in order to sustain a viable population. Modelling tools often rely on the spatial scale of the area as an indicator of conservation value. There has been an increase in emphasis on conserving few large areas of habitat as opposed to many small areas. This idea is often referred to as the "single large or several small", SLOSS debate, and is a highly controversial area among conservation biologists and ecologists. The reasons behind the argument that "larger is better" include the reduction in the negative impacts of patch edge effects, the general idea that species richness increases with habitat area and the ability of larger habitats to support greater populations with lower extinction probabilities. Noss & Cooperrider support the "larger is better" claim and developed a model that implies areas of habitat less than 1000ha are "tiny" and of low conservation value. However, Shwartz suggests that although "larger is better", this does not imply that "small is bad". Shwartz argues that human induced habitat loss leaves no alternative to conserving small areas. Furthermore, he suggests many endangered species which are of high conservation value, may only be restricted to small isolated patches of habitat, and thus would be overlooked if larger areas were given a higher priority. The shift to conserving larger areas is somewhat justified in society by placing more value on larger vertebrate species, which naturally have larger habitat requirements.

Examples of Current Conservation Organizations

The Nature Conservancy

Since its formation in 1951 The Nature Conservancy has slowly developed into one of the world's largest conservation organizations. Currently operating in over 30 countries, across 5 continents world-wide, The Nature Conservancy aims to protect nature and its assets for future generations. The organization purchases land or accepts land donations with the intension of conserving its natural resources. In 1955 The Nature Conservancy purchased its first 60-acre plot near the New York/Connecticut border in the United States of America. Today the Conservancy has expanded to protect over 119 million acres of land, 5,000 river miles as well as participating in over 1000 marine protection programs across the globe . Since its beginnings The Nature Conservancy has understood the benefit in taking a scientific approach towards habitat conservation. For the last decade the organization has been using a collaborative, scientific method known as 'Conservation by Design'. By collecting and analyzing scientific data The Conservancy is able to holistically approach the protection of various ecosystems. This process determines the habitats that need protection, specific elements that should be conserved as well as monitoring progress so more efficient practices can be developed for the future.

The Nature Conservancy currently has a large number of diverse projects in operation. They work with countries around the world to protect forests, river systems, oceans, deserts and grasslands. In all cases the aim is to provide a sustainable environment for both the plant and animal life forms that depend on them as well as all future generations to come. turtles

World Wildlife Fund (WWF)

The World Wildlife Fund (WWF) was first formed in after a group of passionate conservationists signed what is now referred to as the Morges Manifesto. WWF is currently operating in over 100 countries across 5 continents with a current listing of over 5 million supporters. One of the first projects of WWF was assisting in the creation of the Charles

Darwin Research Foundation which aided in the protection of diverse range of unique species existing on the Galápagos' Islands, Ecuador. It was also a WWF grant that helped with the formation of the College of African Wildlife Management in Tanzania which today focuses on teaching a wide range of protected area management skills in areas such as ecology, range management and law enforcement. The WWF has since gone on to aid in the protection of land in Spain, creating the Coto Doñana National Park in order to conserve migratory birds and The Democratic Republic of Congo, home to the world's largest protected wetlands. The WWF also initiated a debt-for-nature concept which allows the country to put funds normally allocated to paying off national debt, into conservation programs that protect its natural landscapes. Countries currently participating include Madagascar, the first country to participate which since 1989 has generated over $US50 million towards preservation, Bolivia, Costa Rica, Ecuador, Gabon, the Philippines and Zambia.

Rare Conservation

Rare has been in operation since 1973 with current global partners in over 50 countries and offices in the United States of America, Mexico, the Philippines, China and Indonesia. Rare focuses on the human activity that threatens biodiversity and habitats such as overfishing and unsustainable agriculture. By engaging local communities and changing behaviour Rare has been able to launch campaigns to protect areas in most need of conservation. The key aspect of Rare's methodology is their "Pride Campaign's". For example, in the Andes in South America, Rare has partnered with 11 different sites with the intention of creating incentives to develop watershed protection practices. In the Southeast Asia's "coral triangle" Rare is training fishers in local communities to better manage the areas around the coral reefs in order to lessen human impact. Such programs last for three years with the aim of changing community attitudes so as to conserve fragile habitats and provide ecological protection for years to come.

WWF Netherlands

WWF Netherlands, along with ARK Nature, Wild Wonders of Europe and Conservation Capital have started the Rewilding Europe project. This project intents to rewild several areas in Europe.

Conservation Biology

Conservation biology is the scientific study of nature and of Earth's biodiversity with the aim of protecting species, their habitats, and ecosystems from excessive rates of extinction and the erosion of biotic interactions. It is an interdisciplinary subject drawing on natural and social sciences, and the practice of natural resource management.

The conservation ethic is based on the findings of conservation biology.

Use of the Term

This term *conservation biology* was introduced as the title of a conference held at the University of California, San Diego in La Jolla, California in 1978 organized by biologists Bruce Wilcox and Michael E. Soulé. The meeting was prompted by the concern among scientists over tropical deforestation, disappearing species, eroding genetic diversity within species. The conference and proceedings that resulted sought to bridge a gap existing at the time between theory in ecology and population biology on the one hand and conservation policy and practice on the other. Conservation biology and the concept of biological diversity (biodiversity) emerged together, helping crystallize the modern era of conservation science and policy. The inherent multidisciplinary basis for conservation biology has led to the development of several new subdisciplines including conservation genetics, conservation social science, conservation behavior and conservation physiology.

Description

The rapid decline of established biological systems around the world means that conservation biology is often referred to as a "Discipline with a deadline". Conservation biology is tied closely to ecology in researching the population ecology (dispersal, migration, demographics, effective population size, inbreeding depression, and minimum population viability) of rare or endangered species. Conservation biology is concerned with phenomena that affect the maintenance, loss, and restoration of biodiversity and the science of sustaining evolutionary processes that engender genetic, population, species, and ecosystem diversity. The concern stems from estimates suggesting that up to 50% of all species on the planet will disappear within the next 50 years, which has contributed to poverty, starvation, and will reset the course of evolution on this planet.

Conservation biologists research and educate on the trends and process of biodiversity loss, species extinctions, and the negative effect these are having on our capabilities to sustain the well-being of human society. Conservation biologists work in the field and office, in government, universities, non-profit organizations and industry. The topics of their research are diverse, because this is an interdisciplinary network with professional alliances in the biological as well as social sciences. Those dedicated to the cause and profession advocate for a global response to the current biodiversity crisis based on morals, ethics, and scientific reason. Organizations and citizens are responding to the biodiversity crisis through conservation action plans that direct research, monitoring, and education programs that engage concerns at local through global scales.

History

The conservation of natural resources is the fundamental problem. Unless we solve that problem, it will avail us little to solve all others.

– *Theodore Roosevelt*

Natural Resource Conservation

Conscious efforts to conserve and protect *global* biodiversity are a recent phenomenon. Natural resource conservation, however, has a history that extends prior to the age of conservation. Resource ethics grew out of necessity through direct relations with nature. Regulation or communal restraint became necessary to prevent selfish motives from taking more than could be locally sustained, therefore compromising the long-term supply for the rest of the community. This social dilemma with respect to natural resource management is often called the "Tragedy of the Commons".

From this principle, conservation biologists can trace communal resource based ethics throughout cultures as a solution to communal resource conflict. For example, the Alaskan Tlingit peoples and the Haida of the Pacific Northwest had resource boundaries, rules, and restrictions among clans with respect to the fishing of sockeye salmon. These rules were guided by clan elders who knew lifelong details of each river and stream they managed. There are numerous examples in history where cultures have followed rules, rituals, and organized practice with respect to communal natural resource management.

Conservation ethics are also found in early religious and philosophical writings. There are examples in the Tao, Shinto, Hindu, Islamic and Buddhist traditions. In Greek philosophy, Plato lamented about pasture land degradation: "What is left now is, so to say, the skeleton of a body wasted by disease; the rich, soft soil has been carried off and only the bare framework of the district left." In the bible, through Moses, God commanded to let the land rest from cultivation every seventh year. Before the 18th century, however, much of European culture considered it a pagan view to admire nature. Wilderness was denigrated while agricultural development was praised. However, as early as AD 680 a wildlife sanctuary was founded on the Farne Islands by St Cuthbert in response to his religious beliefs.

Early naturalists

Natural history was a major preoccupation in the 18th century, with grand expeditions and the opening of popular public displays in Europe and North America. By 1900 there were 150 natural history museums in Germany, 250 in Great Britain, 250 in the United States, and 300 in France. Preservationist or conservationist sentiments are a development of the late 18th to early 20th centuries.

Before Charles Darwin set sail on the HMS Beagle, most people in the world, including Darwin, believed in special creation and that all species were unchanged. George-Louis Leclerc was one of the first naturalist that questioned this belief. He proposed in his 44 volume natural history book that species evolve due to environmental influences. Erasmus Darwin was also a naturalist who also suggested that species evolved. Erasmus Darwin noted that some species have vestigial structures which are anatomical structures that have no apparent function in the species currently but would have been useful for the species' ancestors. The thinking of these early 18th century naturalist helped to change the mindset and thinking of the early 19th century naturalist.

White gyrfalcons drawn by John James Audubon

By the early 19th century biogeography was ignited through the efforts of Alexander von Humboldt, Charles Lyell and Charles Darwin. The 19th-century fascination with natural history engendered a fervor to be the first to collect rare specimens with the goal of doing so before they became extinct by other such collectors. Although the work of many 18th and 19th century naturalists were to inspire nature enthusiasts and conservation organizations, their writings, by modern standards, showed insensitivity towards conservation as they would kill hundreds of specimens for their collections.

Conservation Movement

The modern roots of conservation biology can be found in the late 18th-century Enlightenment period particularly in England and Scotland. A number of thinkers, among them notably Lord Monboddo, described the importance of "preserving nature"; much of this early emphasis had its origins in Christian theology.

Scientific conservation principles were first practically applied to the forests of British India. The conservation ethic that began to evolve included three core principles: that human activity damaged the environment, that there was a civic duty to maintain the environment for future generations, and that scientific, empirically based methods should be applied to ensure this duty was carried out. Sir James Ranald Martin was prominent in promoting this ideology, publishing many medico-topographical reports that demonstrated the scale of damage wrought through large-scale deforestation and desiccation, and lobbying extensively for the institutionalization of forest conservation activities in British India through the establishment of Forest Departments.

The Madras Board of Revenue started local conservation efforts in 1842, headed by Alexander Gibson, a professional botanist who systematically adopted a forest conservation program based on scientific principles. This was the first case of state conservation management of forests in the world. Governor-General Lord Dalhousie introduced the first permanent and large-scale forest conservation program in the world in 1855, a model that soon spread to other colonies, as well the United States, where Yellowstone National Park was opened in 1872 as the world's first national park.

The term *conservation* came into widespread use in the late 19th century and referred to the management, mainly for economic reasons, of such natural resources as timber, fish, game, topsoil, pastureland, and minerals. In addition it referred to the preservation of forests (forestry), wildlife (wildlife refuge), parkland, wilderness, and watersheds. This period also saw the passage of the first conservation legislation and the establishment of the first nature conservation societies. The Sea Birds Preservation Act of 1869 was passed in Britain as the first nature protection law in the world after extensive lobbying from the Association for the Protection of Seabirds and the respected ornithologist Alfred Newton. Newton was also instrumental in the passage of the first Game laws from 1872, which protected animals during their breeding season so as to prevent the stock from being brought close to extinction.

One of the first conservation societies was the Royal Society for the Protection of Birds, founded in 1889 in Manchester as a protest group campaigning against the use of great crested grebe and kittiwake skins and feathers in fur clothing. Originally known as "the Plumage League". the group gained popularity and eventually amalgamated with the Fur and Feather League in Croydon, and formed the RSPB. The National Trust formed in 1895 with the manifesto to "...promote the permanent preservation, for the benefit of the nation, of lands, ...to preserve (so far practicable) their natural aspect."

In the United States, the Forest Reserve Act of 1891 gave the President power to set aside forest reserves from the land in the public domain. John Muir founded the Sierra Club in 1892, and the New York Zoological Society was set up in 1895. A series of national forests and preserves were established by Theodore Roosevelt from 1901 to 1909. The 1916 National Parks Act, included a 'use without impairment' clause, sought by John Muir, which eventually resulted in the removal of a proposal to build a dam in Dinosaur National Monument in 1959.

In the 20th century, Canadian civil servants, including Charles Gordon Hewitt and James Harkin spearheaded the movement toward wildlife conservation.

Roosevelt and Muir on Glacier Point in Yosemite National Park.

Global Conservation Efforts

In the mid-20th century, efforts arose to target individual species for conservation, notably efforts in big cat conservation in South America led by the New York Zoological Society. In the early 20th century the New York Zoological Society was instrumental in developing concepts of establishing preserves for particular species and conducting the necessary conservation studies to determine the suitability of locations that are most appropriate as conservation priorities; the work of Henry Fairfield Osborn Jr., Carl E. Akeley, Archie Carr and Archie Carr III is notable in this era. Akeley for example, having led expeditions to the Virunga Mountains and observed the mountain gorilla in the wild, became convinced that the species and the area were conservation priorities. He was instrumental in persuading Albert I of Belgium to act in defense of the mountain gorilla and establish Albert National Park (since renamed Virunga National Park) in what is now Democratic Republic of Congo.

By the 1970s, led primarily by work in the United States under the Endangered Species Act along with the Species at Risk Act (SARA) of Canada, Biodiversity Action Plans developed in Australia, Sweden, the United Kingdom, hundreds of species specific protection plans ensued. Notably the United Nations acted to conserve sites of outstanding cultural or natural importance to the common heritage of mankind. The programme was adopted by the General Conference of UNESCO in 1972. As of 2006, a total of 830 sites are listed: 644 cultural, 162 natural. The first country to pursue aggressive biological conservation through national legislation was the United States, which passed back to back legislation in the Endangered Species Act (1966) and National Environmental Policy Act (1970), which together injected major funding and protection measures to large-scale habitat protection and threatened species research. Other conservation developments, however, have taken hold throughout the world. India, for example, passed the Wildlife Protection Act of 1972.

In 1980 a significant development was the emergence of the urban conservation movement. A local organization was established in Birmingham, UK, a development followed in rapid succession in cities across the UK, then overseas. Although perceived as a grassroots movement, its early development was driven by academic research into urban wildlife. Initially perceived as radical, the movement's view of conservation being inextricably linked with other human activity has now become mainstream in conservation thought. Considerable research effort is now directed at urban conservation biology. The Society for Conservation Biology originated in 1985.

By 1992 most of the countries of the world had become committed to the principles of conservation of biological diversity with the Convention on Biological Diversity; subsequently many countries began programmes of Biodiversity Action Plans to identify and conserve threatened species within their borders, as well as protect associated habitats. The late 1990s saw increasing professionalism in the sector, with the maturing of organisations such as the Institute of Ecology and Environmental Management and the Society for the Environment.

Since 2000 the concept of landscape scale conservation has risen to prominence, with less emphasis being given to single-species or even single-habitat focused actions. Instead an ecosystem approach is advocated by most mainstream conservationists, although concerns have been expressed by those working to protect some high-profile species.

Ecology has clarified the workings of the biosphere; i.e., the complex interrelationships among humans, other species, and the physical environment. The burgeoning human population and associated agriculture, industry, and the ensuing pollution, have demonstrated how easily ecological relationships can be disrupted.

> “ The last word in ignorance is the man who says of an animal or plant: “What good is it?” If the land mechanism as a whole is good, then every part is good, whether we understand it or not. If the biota, in the course of aeons, has built something we like but do not understand, then who but a fool would discard seemingly useless parts? To keep every cog and wheel is the first precaution of intelligent tinkering. ”
>
> —*Aldo Leopold, A Sand County Almanac*

Concepts and Foundations

Measuring Extinction Rates

Extinction rates are measured in a variety of ways. Conservation biologists measure and apply statistical measures of fossil records, rates of habitat loss, and a multitude of other variables such as loss of biodiversity as a function of the rate of habitat loss and site occupancy to obtain such estimates. *The Theory of Island Biogeography* is possibly the most significant contribution toward the scientific understanding of both the process and how to measure the rate of species extinction. The current background extinction rate is estimated to be one species every few years.

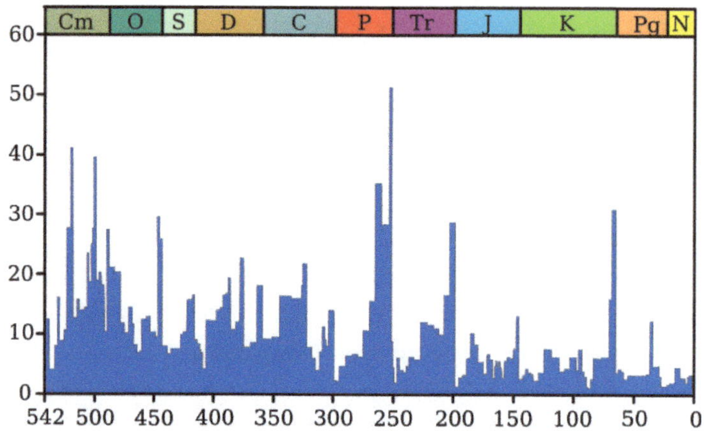

The blue graph shows the apparent *percentage* (not the absolute number) of marine animal genera becoming extinct during any given time interval. It does not represent all marine species, just those that are readily fossilized. The labels of the "Big Five" extinction events are clickable hyperlinks

The measure of ongoing species loss is made more complex by the fact that most of the Earth's species have not been described or evaluated. Estimates vary greatly on how many species actually exist (estimated range: 3,600,000-111,700,000) to how many have received a species binomial (estimated range: 1.5-8 million). Less than 1% of all species that have been described have been studied beyond simply noting its existence. From these figures, the IUCN reports that 23% of vertebrates, 5% of invertebrates and 70% of plants that have been evaluated are designated as endangered or threatened. Better knowledge is being constructed by The Plant List for actual numbers of species.

Systematic Conservation Planning

Systematic conservation planning is an effective way to seek and identify efficient and effective types of reserve design to capture or sustain the highest priority biodiversity values and to work with communities in support of local ecosystems. Margules and Pressey identify six interlinked stages in the systematic planning approach:

1. Compile data on the biodiversity of the planning region
2. Identify conservation goals for the planning region
3. Review existing conservation areas
4. Select additional conservation areas

5. Implement conservation actions
6. Maintain the required values of conservation areas

Conservation biologists regularly prepare detailed conservation plans for grant proposals or to effectively coordinate their plan of action and to identify best management practices (e.g.). Systematic strategies generally employ the services of Geographic Information Systems to assist in the decision making process.

Conservation Physiology: a Mechanistic Approach to Conservation

Conservation Physiology was defined by Steven J. Cooke and colleagues as: 'An integrative scientific discipline applying physiological concepts, tools, and knowledge to characterizing biological diversity and its ecological implications; understanding and predicting how organisms, populations, and ecosystems respond to environmental change and stressors; and solving conservation problems across the broad range of taxa (i.e. including microbes, plants, and animals). Physiology is considered in the broadest possible terms to include functional and mechanistic responses at all scales, and conservation includes the development and refinement of strategies to rebuild populations, restore ecosystems, inform conservation policy, generate decision-support tools, and manage natural resources.' Conservation physiology is particularly relevant to practitioners in that it has the potential to generate cause-and-effect relationships and reveal the factors that contribute to population declines.

Conservation Biology as a Profession

The Society for Conservation Biology is a global community of conservation professionals dedicated to advancing the science and practice of conserving biodiversity. Conservation biology as a discipline reaches beyond biology, into subjects such as philosophy, law, economics, humanities, arts, anthropology, and education. Within biology, conservation genetics and evolution are immense fields unto themselves, but these disciplines are of prime importance to the practice and profession of conservation biology.

Is conservation biology an objective science when biologists advocate for an inherent value in nature? Do conservationists introduce bias when they support policies using qualitative description, such as habitat *degradation*, or *healthy* ecosystems? As all scientists hold values, so do conservation biologists. Conservation biologists advocate for reasoned and sensible management of natural resources and do so with a disclosed combination of science, reason, logic, and values in their conservation management plans. This sort of advocacy is similar to the medical profession advocating for healthy lifestyle options, both are beneficial to human well-being yet remain scientific in their approach.

There is a movement in conservation biology suggesting a new form of leadership is needed to mobilize conservation biology into a more effective discipline that is able to communicate the full scope of the problem to society at large. The movement proposes an adaptive leadership approach that parallels an adaptive management approach. The concept is based on a new philosophy or leadership theory steering away from historical notions of power, authority, and dominance. Adaptive conservation leadership is reflective and more equitable as it applies to any member of society who can mobilize others toward meaningful change using communication techniques that are inspiring, purposeful, and collegial. Adaptive conservation leadership and mentoring programs are being implemented by conservation biologists through organizations such as the Aldo Leopold Leadership Program

Approaches

Conservation may be classified as either in-situ conservation, which is protecting an endangered species in its natural habitat, or ex-situ conservation, which occurs outside the natural habitat. In-situ conservation involves protecting or restoring the habitat. Ex-situ conservation, on the other hand, involves protection outside of an organism's natural habitat, such as on reservations or in gene banks, in circumstances where viable populations may not be present in the natural habitat.

Also, non-interference may be used, which is termed a preservationist method. Preservationists advocate for giving areas of nature and species a protected existence that halts interference from the humans. In this regard, conservationists differ from preservationists in the social dimension, as conservation biology engages society and seeks equitable solutions for both society and ecosystems. Some preservationists emphasize the potential of biodiversity in a world without humans.

Ethics and Values

Conservation biologists are interdisciplinary researchers that practice ethics in the biological and social sciences. Chan states that conservationists must advocate for biodiversity and can do so in a scientifically ethical manner by not promoting simultaneous advocacy against other competing values. A conservationist researches biodiversity and reasons through a Resource Conservation Ethic , which identify what measures will deliver "the greatest good for the greatest number of people for the longest time."

Some conservation biologists argue that nature has an intrinsic value that is independent of anthropocentric usefulness or utilitarianism. Intrinsic value advocates that a gene, or species, be valued because they have a utility for the ecosystems they sustain. Aldo Leopold was a classical thinker and writer on such conservation ethics whose philosophy, ethics and writings are still valued and revisited by modern conservation biologists. His writing is often required reading for those in the profession.

Conservation Priorities

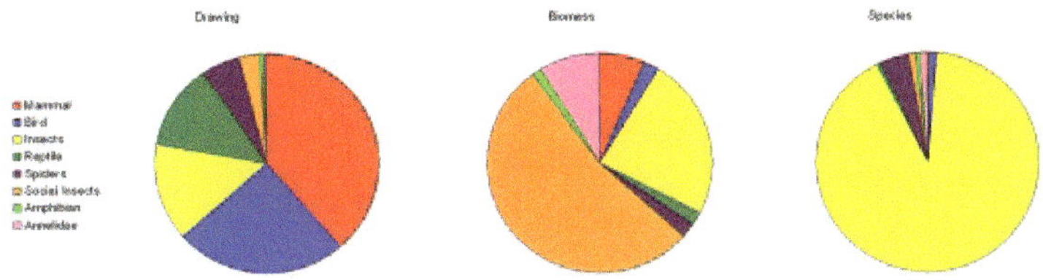

A pie chart image showing the relative biomass representation in a rain forest through a summary of children's perceptions from drawings and artwork (left), through a scientific estimate of actual biomass (middle), and by a measure of biodiversity (right). Notice that the biomass of social insects (middle) far outweighs the number of species (right).

The International Union for the Conservation of Nature (IUCN) has organized a global assortment of scientists and research stations across the planet to monitor the changing state of nature in an effort to tackle the extinction crisis. The IUCN provides annual updates on the status of species conservation through its Red List. The IUCN Red List serves as an international conservation tool to identify those species most in need of conservation attention and by providing a global index on the status of biodiversity. More than the dramatic rates of species loss, however, conservation scientists note that the sixth mass extinction is a biodiversity crisis requiring far more action than a priority focus on rare, endemic or endangered species. Concerns for biodiversity loss covers a broader conservation mandate that looks at ecological processes, such as migration, and a holistic examination of biodiversity at levels beyond the species, including genetic, population and ecosystem diversity. Extensive, systematic, and rapid rates of biodiversity loss threatens the sustained well-being of humanity by limiting supply of ecosystem services that are otherwise regenerated by the complex and evolving holistic network of genetic and ecosystem diversity. While the conservation status of species is employed extensively in conservation management, some scientists highlight that it is the common species that are the primary source of exploitation and habitat alteration by humanity. Moreover, common species are often undervalued despite their role as the primary source of ecosystem services.

While most in the community of conservation science "stress the importance" of sustaining biodiversity, there is debate on how to prioritize genes, species, or ecosystems, which are all components of biodiversity (e.g. Bowen, 1999). While the predominant approach to date has been to focus efforts on endangered species by conserving *biodiversity hotspots*, some scientists (e.g) and conservation organizations, such as the Nature Conservancy, argue that it is more cost-effective, logical, and socially relevant to invest in *biodiversity coldspots*. The costs of discovering, naming, and mapping out the distribution of every species, they argue, is an ill-advised conservation venture. They reason it is better to understand the significance of the ecological roles of species.

Biodiversity hotspots and coldspots are a way of recognizing that the spatial concentration of genes, species, and ecosystems is not uniformly distributed on the Earth's surface. For example, "[...] 44% of all species of vascular plants and 35% of all species in four vertebrate groups are confined to 25 hotspots comprising only 1.4% of the land surface of the Earth."

Those arguing in favor of setting priorities for coldspots point out that there are other measures to consider beyond biodiversity. They point out that emphasizing hotspots downplays the importance of the social and ecological connections to vast areas of the Earth's ecosystems where biomass, not biodiversity, reigns supreme. It is estimated that 36% of the Earth's surface, encompassing 38.9% of the worlds vertebrates, lacks the endemic species to qualify as biodiversity hotspot. Moreover, measures show that maximizing protections for biodiversity does not capture ecosystem services any better than targeting randomly chosen regions. Population level biodiversity (i.e. coldspots) are disappearing at a rate that is ten times that at the species level. The level of importance in addressing biomass versus endemism as a concern for conservation biology is highlighted in literature measuring the level of threat to global ecosystem carbon stocks that do not necessarily reside in areas of endemism. A hotspot priority approach would not invest so heavily in places such as steppes, the Serengeti, the Arctic, or taiga. These areas contribute a great abundance of population (not species) level biodiversity and ecosystem services, including cultural value and planetary nutrient cycling.

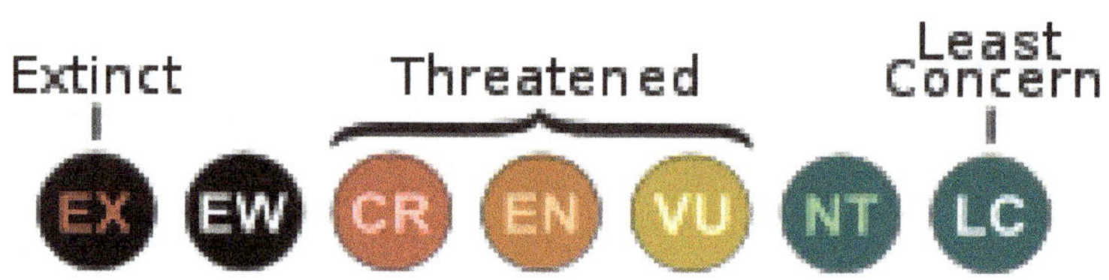

Summary of 2006 IUCN Red List categories.

Those in favor of the hotspot approach point out that species are irreplaceable components of the global ecosystem, they are concentrated in places that are most threatened, and should therefore receive maximal strategic protections. The IUCN Red List categories, which appear on Wikipedia species articles, is an example of the hotspot conservation approach in action; species that are not rare or endemic are listed the least concern and their This is a hotspot approach because the priority is set to target species level concerns over population level or biomass. Species richness and genetic biodiversity contributes to and engenders ecosystem stability, ecosystem processes, evolutionary adaptability, and biomass. Both sides agree, however, that conserving biodiversity is necessary to reduce the extinction rate and identify an inherent value in nature; the debate hinges on how to prioritize limited conservation resources in the most cost-effective way.

Economic Values and Natural Capital

Conservation biologists have started to collaborate with leading global economists to determine how to measure the wealth and services of nature and to make these values apparent in global market transactions. This system of accounting is called *natural capital* and would, for example, register the value of an ecosystem before it is cleared to make way for development. The WWF publishes its *Living Planet Report* and provides a global index of biodiversity by monitoring approximately 5,000 populations in 1,686 species of vertebrate (mammals, birds, fish, reptiles, and amphibians) and report on the trends in much the same way that the stock market is tracked.

Tadrart Acacus desert in western Libya, part of the Sahara.

This method of measuring the global economic benefit of nature has been endorsed by the G8+5 leaders and the European Commission. Nature sustains many ecosystem services that benefit humanity. Many of the earths ecosystem services are public goods without a market and therefore no price or value. When the *stock market* registers a financial crisis, traders on Wall Street are not in the business of trading stocks for much of the planet's living natural capital stored in ecosystems. There is no natural stock market with investment portfolios into sea horses, amphibians, insects, and other creatures that provide a sustainable supply of ecosystem services that are valuable to society. The ecological footprint of society has exceeded the bio-regenerative capacity limits of the planet's ecosystems by about 30 percent, which is the same percentage of vertebrate populations that have registered decline from 1970 through 2005.

The ecological credit crunch is a global challenge. The *Living Planet Report 2008* tells us that more than three-quarters of the world's people live in nations that are ecological debtors – their national consumption has outstripped their country's biocapacity. Thus, most of us are propping up our current lifestyles, and our economic growth, by drawing (and increasingly overdrawing) upon the ecological capital of other parts of the world.

WWF Living Planet Report

The inherent natural economy plays an essential role in sustaining humanity, including the regulation of global atmospheric chemistry, pollinating crops, pest control, cycling soil nutrients, purifying our water supply, supplying medicines and health benefits, and unquantifiable quality of life improvements. There is a relationship, a correlation, between markets and natural capital, and social income inequity and biodiversity loss. This means that there are greater rates of biodiversity loss in places where the inequity of wealth is greatest

Although a direct market comparison of natural capital is likely insufficient in terms of human value, one measure of ecosystem services suggests the contribution amounts to trillions of dollars yearly. For example, one segment of North American forests has been assigned an annual value of 250 billion dollars; as another example, honey-bee pollination is estimated to provide between 10 and 18 billion dollars of value yearly. The value of ecosystem services on one New Zealand island has been imputed to be as great as the GDP of that region. This planetary wealth is being lost at an incredible rate as the demands of human society is exceeding the bio-regenerative capacity of the Earth. While biodiversity and ecosystems are resilient, the danger of losing them is that humans cannot recreate many ecosystem functions through technological innovation.

Strategic Species Concepts

Keystone Species

Some species, called a *keystone species* form a central supporting hub unique to their the ecosystem. The loss of such a species results in a collapse in ecosystem function, as well as the loss of coexisting species. Keystone species are usually predators due to their ability to control the population of prey in their ecosystem. The importance of a keystone species was shown by the extinction of the Steller's sea cow (*Hydrodamalis gigas*) through its interaction with sea otters, sea urchins, and kelp. Kelp beds grow and form nurseries in shallow waters to shelter creatures that support the food chain. Sea urchins feed on kelp, while sea otters feed on sea urchins. With the rapid decline of sea otters due to overhunting, sea urchin populations grazed unrestricted on the kelp beds and the ecosystem collapsed. Left unchecked, the urchins destroyed the shallow water kelp communities that supported the Steller's sea cow's diet and hastened their demise. The sea otter was thought to be a keystone species because the coexistence of many ecological associates in the kelp beds relied upon otters for their survival. However this was later questioned by Turvey and Risley, who showed that hunting alone would have driven the Steller's sea cow extinct.

Indicator Species

An *indicator species* has a narrow set of ecological requirements, therefore they become useful targets for observing the health of an ecosystem. Some animals, such as amphibians with their semi-permeable skin and linkages to wetlands, have an acute sensitivity to environmental harm and thus may serve as a *miner's canary*. Indicator species are monitored in an effort to capture environmental degradation through pollution or some other link to proximate human activities. Monitoring an indicator species is a measure to determine if there is a significant environmental impact that can serve to advise or modify practice, such as through different forest silviculture treatments and management scenarios, or to measure the degree of harm that a pesticide may impart on the health of an ecosystem.

Government regulators, consultants, or NGOs regularly monitor indicator species, however, there are limitations coupled with many practical considerations that must be followed for the approach to be effective. It is generally recommended that multiple indicators (genes, populations, species, communities, and landscape) be monitored for effective conservation measurement that prevents harm to the complex, and often unpredictable, response from ecosystem dynamics (Noss, 1997).

Umbrella and Flagship Species

An example of an *umbrella species* is the monarch butterfly, because of its lengthy migrations and aesthetic value. The monarch migrates across North America, covering multiple ecosystems and so requires a large area to exist. Any protections afforded to the monarch butterfly will at the same time umbrella many other species and habitats. An umbrella species is often used as *flagship species*, which are species, such as the giant panda, the blue whale, the tiger, the mountain gorilla and the monarch butterfly, that capture the public's attention and attract support for conservation measures.

Context and Trends

Conservation biologists study trends and process from the paleontological past to the ecological present as they gain an understanding of the context related to species extinction. It is generally accepted that there have been five major global mass extinctions that register in Earth's history. These include: the Ordovician (440 mya), Devonian (370 mya), Permian–Triassic (245 mya), Triassic–Jurassic (200 mya), and Cretaceous–Paleogene extinction event (66 mya) extinction spasms. Within the last 10,000 years, human influence over the Earth's ecosystems has been so extensive that scientists have difficulty estimating the number of species lost; that is to say the rates of deforestation, reef destruction, wetland draining and other human acts are proceeding much faster than human assessment of species. The latest *Living Planet Report* by the World Wide Fund for Nature estimates that we have exceeded the bio-regenerative capacity of the planet, requiring 1.5 Earths to support the demands placed on our natural resources.

Holocene Extinction

Conservation biologists are dealing with and have published evidence from all corners of the planet indicating that humanity may be causing the sixth and fastest planetary extinction event. It has been suggested that we are living in an era of unprecedented numbers of species extinctions, also known as the Holocene extinction event. The global extinction rate may be approximately 1,000 times higher than the natural background extinction rate. It is estimated that two-thirds of all mammal genera and one-half of all mammal species weighing at least 44 kilograms (97 lb) have gone extinct in the last 50,000 years. The Global Amphibian Assessment reports that amphibians are declining on a global scale faster than any other vertebrate group, with over 32% of all surviving species being threatened with extinction. The surviving populations are in continual decline in 43% of those that are threatened. Since the mid-1980s the actual rates of extinction have exceeded 211 times rates measured from the fossil record. However, "The current amphibian extinction rate may range from 25,039 to 45,474 times the background extinction rate for amphibians." The global extinction trend occurs in every major vertebrate group that is being monitored. For example, 23% of all mammals and 12% of all birds are Red Listed by the International Union for Conservation of Nature (IUCN), meaning they too are threatened with extinction. Even though extinction is natural, the decline in species is happening at such an incredible rate that evolution can simply not match, therefore, leading to the greatest continual mass extinction on Earth. Humans have dominated the planet and our high consumption of resources, along with the pollution generated is affecting the environments in which other species live. There are a wide variety of species that humans are working to protect such as the Hawaiian Crow and the Whooping Crane of Texas. People can also take action on preserving species by advocating and voting for policies that improve climate. The Earth's oceans especially require attention, as climate change has altered pH levels making it inhabitable for organisms with shells, that are dissolving as a result.

An art scape image showing the relative importance of animals in a rain forest through a summary of (a) child's perception compared with (b) a scientific estimate of the importance. The size of the animal represents its importance. The child's mental image places importance on big cats, birds, butterflies, and then reptiles versus the actual dominance of social insects (such as ants).

Status of Oceans and Reefs

Global assessments of coral reefs of the world continue to report drastic and rapid rates of decline. By 2000, 27% of the world's coral reef ecosystems had effectively collapsed. The largest period of decline occurred in a dramatic "bleaching" event in 1998, where approximately 16% of all the coral reefs in the world disappeared in less than a year. *Coral bleaching* is caused by a mixture of environmental stresses, including increases in ocean temperatures and acidity, causing both the release of symbiotic algae and death of corals. Decline and extinction risk in coral reef biodiversity has risen dramatically in the past ten years. The loss of coral reefs, which are predicted to go extinct in the next century, threatens the balance of global biodiversity, will have huge economic impacts, and endangers food security for hundreds of millions of people. Conservation biology plays an important role in international agreements covering the world's oceans (and other issues pertaining to biodiversity).

These predictions will undoubtedly appear extreme, but it is difficult to imagine how such changes will not come to pass without fundamental changes in human behavior.

J.B. Jackson

The oceans are threatened by acidification due to an increase in CO_2 levels. This is a most serious threat to societies relying heavily upon oceanic natural resources. A concern is that the majority of all marine species will not be able to evolve or acclimate in response to the changes in the ocean chemistry.

The prospects of averting mass extinction seems unlikely when "[...] 90% of all of the large (average approximately ≥50 kg), open ocean tuna, billfishes, and sharks in the ocean" are reportedly gone. Given the scientific review of current trends, the ocean is predicted to have few surviving multi-cellular organisms with only microbes left to dominate marine ecosystems.

Groups Other than Vertebrates

Serious concerns also being raised about taxonomic groups that do not receive the same degree of social attention or attract funds as the vertebrates. These include fungal (including lichen-forming species), invertebrate (particularly insect) and plant communities where the vast majority of biodiversity is represented. Conservation of fungi and conservation of insects, in particular, are both of pivotal importance for conservation biology. As mycorrhizal symbionts, and as decomposers and recyclers, fungi are essential for sustainability of forests. The value of insects in the biosphere is enormous because they outnumber all other living groups in measure of species richness. The greatest bulk of biomass on land is found in plants, which is sustained by insect relations. This great ecological value of insects is countered by a society that often reacts negatively toward these aesthetically 'unpleasant' creatures.

One area of concern in the insect world that has caught the public eye is the mysterious case of missing honey bees (*Apis mellifera*). Honey bees provide an indispensable ecological services through their acts of pollination supporting a huge variety of agriculture crops. The use of honey and wax have become vastly used throughout the world. The sudden disappearance of bees leaving empty hives or colony collapse disorder (CCD) is not uncommon. However, in 16-month period from 2006 through 2007, 29% of 577 beekeepers across the United States reported CCD losses in up to 76% of their colonies.

This sudden demographic loss in bee numbers is placing a strain on the agricultural sector. The cause behind the massive declines is puzzling scientists. Pests, pesticides, and global warming are all being considered as possible causes.

Another highlight that links conservation biology to insects, forests, and climate change is the mountain pine beetle (*Dendroctonus ponderosae*) epidemic of British Columbia, Canada, which has infested 470,000 km² (180,000 sq mi) of forested land since 1999. An action plan has been prepared by the Government of British Columbia to address this problem.

This impact [*pine beetle epidemic*] converted the forest from a small net carbon sink to a large net carbon source both during and immediately after the outbreak. In the worst year, the impacts resulting from the beetle outbreak in British Columbia were equivalent to 75% of the average annual direct forest fire emissions from all of Canada during 1959–1999.

—*Kurz et al.*

Conservation Biology of Parasites

A large proportion of parasite species are threatened by extinction. A few of them are being eradicated as pests of humans or domestic animals, however, most of them are harmless. Threats include the decline or fragmentation of host populations, or the extinction of host species.

Threats to Biodiversity

Today, many threats to Biodiversity exist. An acronym that can be used to express the top threats of present-day H.I.P.P.O stands for Habitat Loss, Invasive Species, Pollution, Human Population, and Overharvesting. The primary threats to biodiversity are habitat destruction (such as deforestation, agricultural expansion, urban development), and overexploitation (such as wildlife trade). Habitat fragmentation also poses challenges, because the global network of protected areas only covers 11.5% of the Earth's surface. A significant consequence of fragmentation and lack of linked protected areas is the reduction of animal migration on a global scale. Considering that billions of tonnes of biomass are responsible for nutrient cycling across the earth, the reduction of migration is a serious matter for conservation biology.

Human activities are associated directly or indirectly with nearly every aspect of the current extinction spasm.

Wake and Vredenburg

However, human activities need not necessarily cause irreparable harm to the biosphere. With conservation management and planning for biodiversity at all levels, from genes to ecosystems, there are examples where humans mutually coexist in a sustainable way with nature. Even with the current threats to biodiversity there are ways we can improve the current condition and start anew.

Many of the threats to biodiversity, including disease and climate change, are reaching inside borders of protected areas, leaving them 'not-so protected' (e.g. Yellowstone National Park). Climate change, for example, is often cited as a serious threat in this regard, because there is a feedback loop between species extinction and the release of carbon dioxide into the atmosphere. Ecosystems store and cycle large amounts of carbon which regulates global conditions. In present day, there have been major climate shifts with temperature changes making survival of some species difficult. The effects of global warming add a catastrophic threat toward a mass extinction of global biological diversity. Conservationists have claimed that not all the species can be saved, and they have to decide which their efforts should be used to protect. This concept is known as the Conservation Triage. The extinction threat is estimated to range from 15 to 37 percent of all species by 2050, or 50 percent of all species over the next 50 years. The current extinction rate is 100-100,000 times more rapid today than the last several billion years.

Conservation Movement

The conservation movement, also known as nature conservation, is a political, environmental and a social movement that seeks to protect natural resources including animal and plant species as well as their habitat for the future.

The early conservation movement included fisheries and wildlife management, water, soil conservation and sustainable forestry. The contemporary conservation movement has broadened from the early movement's emphasis on use of sustainable yield of natural resources and preservation of wilderness areas to include preservation of biodiversity. Some say the conservation movement is part of the broader and more far-reaching environmental movement, while others argue that they differ both in ideology and practice. Chiefly in the United States, conservation is seen as differing from environmentalism in that it aims to preserve natural resources expressly for their continued sustainable use by humans. In other parts of the world conservation is used more broadly to include the setting aside of natural areas and the active protection of wildlife for their inherent value, as much as for any value they may have for humans.

History

Early History

The conservation movement can be traced back to John Evelyn's work *Sylva*, presented as a paper to the Royal Society in 1662. Published as a book two years later, it was one of the most highly influential texts on forestry ever published. Timber resources in

England were becoming dangerously depleted at the time, and Evelyn advocated the importance of conserving the forests by managing the rate of depletion and ensuring that the cut down trees get replenished.

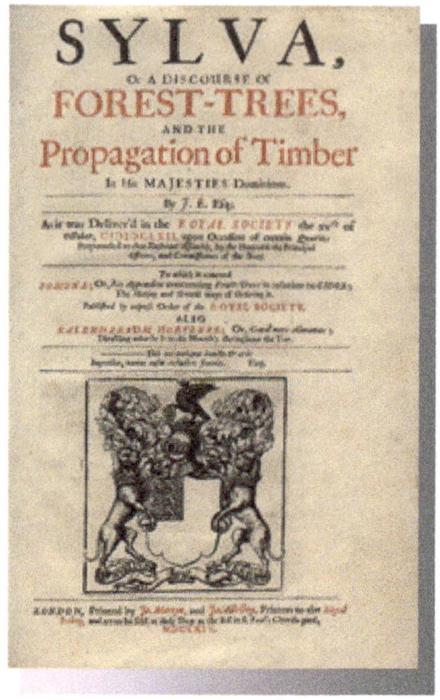
SYLVA,
Or A DISCOURSE Of
FOREST-TREES,
AND THE
Propagation of Timber
In His MAJESTIES Dominions.
By J. E. Esq;

Sylva, or A Discourse of Forest-Trees and the Propagation of Timber in His Majesty's Dominions, title page of the first edition (1664).

The field developed during the 18th century, especially in Prussia and France where scientific forestry methods were developed. These methods were first applied rigorously in British India from the early-19th century. The government was interested in the use of forest produce and began managing the forests with measures to reduce the risk of wildfire in order to protect the "household" of nature, as it was then termed. This early ecological idea was in order to preserve the growth of delicate teak trees, which was an important resource for the Royal Navy. Concerns over teak depletion were raised as early as 1799 and 1805 when the Navy was undergoing a massive expansion during the Napoleonic Wars; this pressure led to the first formal conservation Act, which prohibited the felling of small teak trees. The first forestry officer was appointed in 1806 to regulate and preserve the trees necessary for shipbuilding. This promising start received a setback in the 1820s and 30s, when laissez-faire economics and complaints from private landowners brought these early conservation attempts to an end.

Origins of the Modern Conservation Movement

Conservation was revived in the mid-19th century, with the first practical application of scientific conservation principles to the forests of India. The conservation ethic that began to evolve included three core principles: that human activity damaged the environment, that there was a civic duty to maintain the environment for future generations, and that scientific, empirically based methods should be applied to ensure this duty was carried out. Sir James Ranald Martin was prominent in promoting this ideology, publishing many medico-topographical reports that demonstrated the scale of damage wrought through large-scale deforestation and desiccation, and lobbying extensively for the institutionalization of forest conservation activities in British India through the establishment of Forest Departments. Edward Percy Stebbing warned of desertification of India. The Madras Board of Revenue started local conservation efforts in 1842, headed by Alexander Gibson, a professional botanist who systematically adopted a forest conservation program based on scientific principles. This was the first case of state management of forests in the world.

These local attempts gradually received more attention by the British government as the unregulated felling of trees continued unabated. In 1850, the British Association in Edinburgh formed a committee to study forest destruction at the behest of Dr. Hugh Cleghorn a pioneer in the nascent conservation movement.

He had become interested in forest conservation in Mysore in 1847 and gave several lectures at the Association on the failure of agriculture in India. These lectures influenced the government under Governor-General Lord Dalhousie to introduce the first permanent and large-scale forest conservation program in the world in 1855, a model that soon spread to other colonies, as well the United States. In the same year, Cleghorn organised the Madras Forest Department and in 1860 the Department banned the use shifting cultivation. Cleghorn's 1861 manual, *The forests and gardens of South India*, became the definitive work on the subject and was widely used by forest assistants in the subcontinent. In 1861, the Forest Department extended its remit into the Punjab.

Schlich, in the middle of the seated row, with students from the forestry school at Oxford, on a visit to the forests of Saxony in the year 1892.

Sir Dietrich Brandis, a German forester, joined the British service in 1856 as superintendent of the teak forests of Pegu division in eastern Burma. During that time Burma's teak forests were controlled by militant Karen tribals. He introduced the "taungya" system, in which Karen villagers provided labour for clearing, planting and weeding teak plantations. After seven years in Burma, Brandis was appointed Inspector General of Forests in India, a position he served in for 20 years. He formulated new forest legislation and helped establish research and training institutions. The Imperial Forest School at Dehradun was founded by him.

Germans were prominent in the forestry administration of British India. As well as Brandis, Berthold Ribbentrop and Sir William P.D. Schlich brought new methods to Indian conservation, the latter becoming the Inspector-General in 1883 after Brandis stepped down. Schlich helped to establish the journal *Indian Forester* in 1874, and became the founding director of the first forestry school in England at Cooper's Hill in 1885. He authored the five-volume *Manual of Forestry* (1889–96) on silviculture, forest management, forest protection, and forest utilisation, which became the standard and enduring textbook for forestry students.

Conservation in the United States

F. V. Hayden's map of Yellowstone National Park, 1871.

The American movement received its inspiration from 19th century works that exalted the inherent value of nature, quite apart from human usage. Author Henry David Thoreau (1817-1862) made key philosophical contributions that exalted nature. Thoreau was interested in peoples' relationship with nature and studied this by living close to nature in a simple life. He published his experiences in the book *Walden,* which argued that people should become intimately close with nature. The ideas of Sir Brandis, Sir William P.D. Schlich and Carl A. Schenck were also very influential - Gifford Pinchot, the first chief of the USDA Forest Service, relied heavily upon Brandis' advice for introducing professional forest management in the U.S. and on how to structure the Forest Service.

Both Conservationists and Preservationists appeared in political debates during the Progressive Era (the 1890s—early 1920s). There were three main positions. The laissez-faire position held that owners of private property—including lumber and mining companies, should be allowed to do anything they wished for their property.

The conservationists, led by future President Theodore Roosevelt and his close ally George Bird Grinnell, were motivated by the wanton waste that was taking place at the hand of market forces, including logging and hunting. This practice resulted in placing a large number of North American game species on the edge of extinction. Roosevelt recognized that the laissez-faire approach of the U.S. Government was too wasteful and inefficient. In any case, they noted, most of the natural resources in the western states were already owned by the federal government. The best course of action, they argued, was a long-term plan devised by national experts to maximize the long-term economic benefits of natural resources. To accomplish the mission, Roosevelt and Grinnell formed the Boone and Crockett Club in 1887. The Club was made up of the best minds and influential men of the day. The Boone and Crockett Club's contingency of conservationists, scientists, politicians, and intellectuals became Roosevelt's closest advisers during his march to preserve wildlife and habitat across North America. Preservationists, led by John Muir (1838–1914), argued that the conservation policies were not strong enough to protect the interest of the natural world because they continued to focus on the natural world as a source of economic production.

The debate between conservation and preservation reached its peak in the public debates over the construction of California's Hetch Hetchy dam in Yosemite National Park which supplies the water supply of San Francisco. Muir, leading the Sierra Club, declared that the valley must be preserved for the sake of its beauty: "No holier temple has ever been consecrated by the heart of man."

President Roosevelt put conservationist issue high on the national agenda. He worked with all the major figures of the movement, especially his chief advisor on the matter, Gifford Pinchot and was deeply committed to conserving natural resources. He encouraged the Newlands Reclamation Act of 1902 to promote federal construction of dams to irrigate small farms and placed 230 million acres (360,000 mi^2 or 930,000 km^2) under federal protection. Roosevelt set aside more federal land for national parks and nature preserves than all of his predecessors combined.

Roosevelt established the United States Forest Service, signed into law the creation of five national parks, and signed the year 1906 Antiquities Act, under which he proclaimed 18 new national monuments. He also established the first 51 bird reserves, four game preserves, and 150 national forests, including Shoshone National Forest, the nation's first. The area of the United States that he placed under public protection totals approximately 230,000,000 acres (930,000 km^2).

Gifford Pinchot had been appointed by McKinley as chief of Division of Forestry in the Department of Agriculture. In 1905, his department gained control of the national forest reserves. Pinchot promoted private use (for a fee) under federal supervision. In

1907, Roosevelt designated 16 million acres (65,000 km²) of new national forests just minutes before a deadline.

A PRACTICAL FORESTER
(A subject that had attention all through Mr. Roosevelt's Presidency.)
From the *Pioneer Press* (St. Paul)

In May 1908, Roosevelt sponsored the Conference of Governors held in the White House, with a focus on natural resources and their most efficient use. Roosevelt delivered the opening address: "Conservation as a National Duty.".

In 1903 Roosevelt toured the Yosemite Valley with John Muir, who had a very different view of conservation, and tried to minimize commercial use of water resources and forests. Working through the Sierra Club he founded, Muir succeeded in 1905 in having Congress transfer the Mariposa Grove and Yosemite Valley to the federal government. While Muir wanted nature preserved for its own sake, Roosevelt subscribed to Pinchot's formulation, "to make the forest produce the largest amount of whatever crop or service will be most useful, and keep on producing it for generation after generation of men and trees."

Theodore Roosevelt's view on conservationism remained dominant for decades; - Franklin D. Roosevelt authorised the building of many large-scale dams and water projects, as well as the expansion of the National Forest System to buy out sub-marginal farms. In 1937, the Pittman–Robertson Federal Aid in Wildlife Restoration Act was signed into law, providing funding for state agencies to carry out their conservation efforts.

Since 1970

Environmental reemerged on the national agenda in 1970, with Republican Richard Nixon playing a major role, especially with his creation of the Environmental Protection Agency. The debates over the public lands and environmental politics played a supporting role in the decline of liberalism and the rise of modern environmentalism.

Although Americans consistently rank environmental issues as "important", polling data indicates that in the voting booth voters rank the environmental issues low relative to other political concerns.

The growth of the Republican party's political power in the inland West (apart from the Pacific coast) was facilitated by the rise of popular opposition to public lands reform. Successful Democrats in the inland West and Alaska typically take more conservative positions on environmental issues than Democrats from the Coastal states. Conservatives drew on new organizational networks of think tanks, industry groups, and citizen-oriented organizations, and they began to deploy new strategies that affirmed the rights of individuals to their property, protection of extraction rights, to hunt and recreate, and to pursue happiness unencumbered by the federal government at the expense of resource conservation.

Areas of Concern

Deforestation and overpopulation are issues affecting all regions of the world. The consequent destruction of wildlife habitat has prompted the creation of conservation groups in other countries, some founded by local hunters who have witnessed declining wildlife populations first hand. Also, it was highly important for the conservation movement to solve problems of living conditions in the cities and the overpopulation of such places.

Boreal Forest and the Arctic

The idea of incentive conservation is a modern one but its practice has clearly defended some of the sub Arctic wildernesses and the wildlife in those regions for thousands of years, especially by indigenous peoples such as the Evenk, Yakut, Sami, Inuit and Cree. The fur trade and hunting by these peoples have preserved these regions for thousands of years. Ironically, the pressure now upon them comes from non-renewable resources such as oil, sometimes to make synthetic clothing which is advocated as a humane substitute for fur. Similarly, in the case of the beaver, hunting and fur trade were thought to bring about the animal's demise, when in fact they were an integral part of its conservation. For many years children's books stated and still do, that the decline in the beaver population was due to the fur trade. In reality however, the decline in beaver numbers was because of habitat destruction and deforestation, as well as its continued persecution as a pest (it causes flooding). In Cree lands however, where the population valued the animal for meat and fur, it continued to thrive. The Inuit defend their relationship with the seal in response to outside critics.

Latin America (Bolivia)

The Izoceño-Guaraní of Santa Cruz Department, Bolivia is a tribe of hunters who were influential in establishing the Capitania del Alto y Bajo Isoso (CABI). CABI promotes economic growth and survival of the Izoceno people while discouraging the rapid destruction of habitat within Bolivia's Gran Chaco. They are responsible for the creation of the 34,000 square kilometre Kaa-Iya del Gran Chaco National Park and Integrated Management Area (KINP). The KINP protects the most biodiverse portion of the Gran Chaco, an ecoregion shared with Argentina, Paraguay and Brazil. In 1996, the Wildlife Conservation Society joined forces with CABI to institute wildlife and hunting monitoring programs in 23 Izoceño communities. The partnership combines traditional beliefs and local knowledge with the political and administrative tools needed to effectively manage habitats. The programs rely solely on voluntary participation by local hunters who perform self-monitoring techniques and keep records of their hunts. The information obtained by the hunters participating in the program has provided CABI with important data required to make educated decisions about the use of the land. Hunters have been willing participants in this program because of pride in their traditional activities, encouragement by their communities and expectations of benefits to the area.

Africa (Botswana)

In order to discourage illegal South African hunting parties and ensure future local use and sustainability, indigenous hunters in Botswana began lobbying for and implementing conservation practices in the 1960s. The Fauna Preservation Society of Ngamiland (FPS) was formed in 1962 by the husband and wife team: Robert Kay and June Kay, environmentalists working in conjunction with the Batawana tribes to preserve wildlife habitat.

The FPS promotes habitat conservation and provides local education for preservation of wildlife. Conservation initiatives were met with strong opposition from the Botswana government because of the monies tied to big-game hunting. In 1963, BaTawanga Chiefs and tribal hunter/adventurers in conjunction with the FPS founded Moremi National Park and Wildlife Refuge, the first area to be set aside by tribal people rather than governmental forces. Moremi National Park is home to a variety of wildlife, including lions, giraffes, elephants, buffalo, zebra, cheetahs and antelope, and covers an area of 3,000 square kilometers. Most of the groups involved with establishing this protected land were involved with hunting and were motivated by their personal observations of declining wildlife and habitat.

References

- Kunin, W.E.; Gaston, Kevin, eds. (31 December 1996). The Biology of Rarity: Causes and consequences of rare—common differences. ISBN 978-0412633805. Retrieved 26 May 2015.
- Stearns, Beverly Peterson; Stearns, S. C.; Stearns, Stephen C. (2000). Watching, from the Edge of Extinction. Yale University Press. p. 1921. ISBN 978-0-300-08469-6. Retrieved 2014-12-27.
- G. Miller; Scott Spoolman (2012). Environmental Science - Biodiversity Is a Crucial Part of the Earth's Natural Capital. Cengage Learning. p. 62. ISBN 1-133-70787-4. Retrieved 2014-12-27.
- Hamilton Raven, Peter; Brooks Johnson, George (2002). Biology. McGraw-Hill Education. p. 68. ISBN 978-0-07-112261-0. Retrieved 2013-07-07.
- Soulé, Michael E. (1980). Conservation biology: an evolutionary-ecological perspective. Sinauer

Associates. ISBN 978-0-87893-800-1.

- Tor-Björn Larsson (2001). Biodiversity evaluation tools for European forests. Wiley-Blackwell. p. 178. ISBN 978-87-16-16434-6. Retrieved 28 June 2011.
- Davis. Intro To Env Engg (Sie), 4E. McGraw-Hill Education (India) Pvt Ltd. pp. 4–. ISBN 978-0-07-067117-1. Retrieved 28 June 2011.
- Sala, Osvaldo E.; Meyerson, Laura A.; Parmesan, Camille (26 January 2009). Biodiversity change and human health: from ecosystem services to spread of disease. Island Press. pp. 3–5. ISBN 978-1-59726-497-6. Retrieved 28 June 2011.
- D. L. Hawksworth (1996). Biodiversity: measurement and estimation. Springer. p. 6. ISBN 978-0-412-75220-9. Retrieved 28 June 2011.
- Gaston, Kevin J.; Spicer, John I. (13 February 2004). Biodiversity: An Introduction. Wiley. ISBN 978-1-4051-1857-6.
- Morand, Serge; Krasnov, Boris R. (1 September 2010). The Biogeography of Host-Parasite Interactions. Oxford University Press. pp. 93–94. ISBN 978-0-19-956135-3. Retrieved 28 June 2011.
- Chivian, Eric, ed. (15 May 2008). Sustaining Life: How Human Health Depends on Biodiversity. OUP USA. ISBN 978-0-19-517509-7.
- Corvalán, Carlos; Hales, Simon (2005). Ecosystems and Human Well-being: Health Synthesis. World Health Organization. pp. 28–. ISBN 978-92-4-156309-3.
- The World Bank (30 June 2010). Water and Development: An Evaluation of World Bank Support, 1997-2007. World Bank Publications. pp. 79–. ISBN 978-0-8213-8394-0.
- Ehrlich, Paul R.; Ehrlich, Anne H. (1983). Extinction: The Causes and Consequences of the Disappearance of Species. Ballantine Books. ISBN 978-0-345-33094-9.
- Moulton, Michael P.; Sanderson, James (1 September 1998). Wildlife Issues in a Changing World. CRC-Press. ISBN 978-1-56670-351-2.
- Jude, David auth., ed. by M. Munawar (1995). The lake Huron ecosystem: ecology, fisheries and management. Amsterdam: S.P.B. Academic Publishing. ISBN 90-5103-117-3.
- Lovejoy, Thomas E.; Hannah, Lee Jay (2005). Climate Change and Biodiversity. New Haven: Yale University Press. pp. 41–55. ISBN 978-0-300-10425-7.
- Davis, Peter (1996). Museums and the natural environment: the role of natural history museums in biological conservation. Leicester University Press. ISBN 978-0-7185-1548-5.
- Dyke, Fred Van (29 February 2008). Conservation Biology: Foundations, Concepts, Applications. Springer Science & Business Media. ISBN 978-1-4020-6890-4.
- Soulé, Michael E. (1 January 1986). Conservation Biology: The Science of Scarcity and Diversity. Sinauer Associates. ISBN 978-0-87893-794-3.

Permissions

All chapters in this book are published with permission under the Creative Commons Attribution Share Alike License or equivalent. Every chapter published in this book has been scrutinized by our experts. Their significance has been extensively debated. The topics covered herein carry significant information for a comprehensive understanding. They may even be implemented as practical applications or may be referred to as a beginning point for further studies.

We would like to thank the editorial team for lending their expertise to make the book truly unique. They have played a crucial role in the development of this book. Without their invaluable contributions this book wouldn't have been possible. They have made vital efforts to compile up to date information on the varied aspects of this subject to make this book a valuable addition to the collection of many professionals and students.

This book was conceptualized with the vision of imparting up-to-date and integrated information in this field. To ensure the same, a matchless editorial board was set up. Every individual on the board went through rigorous rounds of assessment to prove their worth. After which they invested a large part of their time researching and compiling the most relevant data for our readers.

The editorial board has been involved in producing this book since its inception. They have spent rigorous hours researching and exploring the diverse topics which have resulted in the successful publishing of this book. They have passed on their knowledge of decades through this book. To expedite this challenging task, the publisher supported the team at every step. A small team of assistant editors was also appointed to further simplify the editing procedure and attain best results for the readers.

Apart from the editorial board, the designing team has also invested a significant amount of their time in understanding the subject and creating the most relevant covers. They scrutinized every image to scout for the most suitable representation of the subject and create an appropriate cover for the book.

The publishing team has been an ardent support to the editorial, designing and production team. Their endless efforts to recruit the best for this project, has resulted in the accomplishment of this book. They are a veteran in the field of academics and their pool of knowledge is as vast as their experience in printing. Their expertise and guidance has proved useful at every step. Their uncompromising quality standards have made this book an exceptional effort. Their encouragement from time to time has been an inspiration for everyone.

The publisher and the editorial board hope that this book will prove to be a valuable piece of knowledge for students, practitioners and scholars across the globe.

Index